Graduate Texts in Mathematics 237

Graduate Texts in Mathematics

1 TAKEUTI/ZARING. Introduction to Axiomatic Set Theory. 2nd ed.
2 OXTOBY. Measure and Category. 2nd ed.
3 SCHAEFER. Topological Vector Spaces. 2nd ed.
4 HILTON/STAMMBACH. A Course in Homological Algebra. 2nd ed.
5 MAC LANE. Categories for the Working Mathematician. 2nd ed.
6 HUGHES/PIPER. Projective Planes.
7 J.-P. SERRE. A Course in Arithmetic.
8 TAKEUTI/ZARING. Axiomatic Set Theory.
9 HUMPHREYS. Introduction to Lie Algebras and Representation Theory.
10 COHEN. A Course in Simple Homotopy Theory.
11 CONWAY. Functions of One Complex Variable I. 2nd ed.
12 BEALS. Advanced Mathematical Analysis.
13 ANDERSON/FULLER. Rings and Categories of Modules. 2nd ed.
14 GOLUBITSKY/GUILLEMIN. Stable Mappings and Their Singularities.
15 BERBERIAN. Lectures in Functional Analysis and Operator Theory.
16 WINTER. The Structure of Fields.
17 ROSENBLATT. Random Processes. 2nd ed.
18 HALMOS. Measure Theory.
19 HALMOS. A Hilbert Space Problem Book. 2nd ed.
20 HUSEMOLLER. Fibre Bundles. 3rd ed.
21 HUMPHREYS. Linear Algebraic Groups.
22 BARNES/MACK. An Algebraic Introduction to Mathematical Logic.
23 GREUB. Linear Algebra. 4th ed.
24 HOLMES. Geometric Functional Analysis and Its Applications.
25 HEWITT/STROMBERG. Real and Abstract Analysis.
26 MANES. Algebraic Theories.
27 KELLEY. General Topology.
28 ZARISKI/SAMUEL. Commutative Algebra. Vol. I.
29 ZARISKI/SAMUEL. Commutative Algebra. Vol. II.
30 JACOBSON. Lectures in Abstract Algebra I. Basic Concepts.
31 JACOBSON. Lectures in Abstract Algebra II. Linear Algebra.
32 JACOBSON. Lectures in Abstract Algebra III. Theory of Fields and Galois Theory.
33 HIRSCH. Differential Topology.
34 SPITZER. Principles of Random Walk. 2nd ed.
35 ALEXANDER/WERMER. Several Complex Variables and Banach Algebras. 3rd ed.
36 KELLEY/NAMIOKA et al. Linear Topological Spaces.
37 MONK. Mathematical Logic.
38 GRAUERT/FRITZSCHE. Several Complex Variables.
39 ARVESON. An Invitation to C^*-Algebras.
40 KEMENY/SNELL/KNAPP. Denumerable Markov Chains. 2nd ed.
41 APOSTOL. Modular Functions and Dirichlet Series in Number Theory. 2nd ed.
42 J.-P. SERRE. Linear Representations of Finite Groups.
43 GILLMAN/JERISON. Rings of Continuous Functions.
44 KENDIG. Elementary Algebraic Geometry.
45 LOÈVE. Probability Theory I. 4th ed.
46 LOÈVE. Probability Theory II. 4th ed.
47 MOISE. Geometric Topology in Dimensions 2 and 3.
48 SACHS/WU. General Relativity for Mathematicians.
49 GRUENBERG/WEIR. Linear Geometry. 2nd ed.
50 EDWARDS. Fermat's Last Theorem.
51 KLINGENBERG. A Course in Differential Geometry.
52 HARTSHORNE. Algebraic Geometry.
53 MANIN. A Course in Mathematical Logic.
54 GRAVER/WATKINS. Combinatorics with Emphasis on the Theory of Graphs.
55 BROWN/PEARCY. Introduction to Operator Theory I: Elements of Functional Analysis.
56 MASSEY. Algebraic Topology: An Introduction.
57 CROWELL/FOX. Introduction to Knot Theory.
58 KOBLITZ. p-adic Numbers, p-adic Analysis, and Zeta-Functions. 2nd ed.
59 LANG. Cyclotomic Fields.
60 ARNOLD. Mathematical Methods in Classical Mechanics. 2nd ed.
61 WHITEHEAD. Elements of Homotopy Theory.
62 KARGAPOLOV/MERIZJAKOV. Fundamentals of the Theory of Groups.
63 BOLLOBAS. Graph Theory.

(*continued after index*)

Rubén A. Martínez-Avendaño
Peter Rosenthal

An Introduction to Operators on the Hardy-Hilbert Space

Rubén A. Martínez-Avendaño
Centro de Investigación en
Matemáticas
Universidad Autónoma del Estado
de Hidalgo, Pachuca, Hidalgo 42184
Mexico
rubenma@uaeh.edu.mx

Peter Rosenthal
Department of Mathematics
University of Toronto,
Toronto, ON M5S 2E4
Canada
rosent@math.toronto.edu

Mathematics Subject Classification (2000): 46-xx, 47-xx, 30-xx

ISBN-13: 978-1-4419-2253-3
e-ISBN-10: 0-387-48578-3
e-ISBN-13: 978-0-387-48578-2

Printed on acid-free paper.

9 8 7 6 5 4 3 2 1

springer.com

To

Rubén, Amanda,
Miguel, Lorena,
Sofía, Andrea,
José Miguel

Carol, Alan,
Jeffrey, Michael,
Daniel, Esther,
Jeremy, Aaron,
Margaret

Preface

The great mathematician G.H. Hardy told us that "Beauty is the first test: there is no permanent place in the world for ugly mathematics" (see [24, p. 85]). It is clear why Hardy loved complex analysis: it is a very beautiful part of classical mathematics. The theory of Hilbert spaces and of operators on them is almost as classical and is perhaps as beautiful as complex analysis. The study of the Hardy–Hilbert space (a Hilbert space whose elements are analytic functions), and of operators on that space, combines these two subjects. The interplay produces a number of extraordinarily elegant results.

For example, very elementary concepts from Hilbert space provide simple proofs of the Poisson integral (Theorem 1.1.21 below) and Cauchy integral (Theorem 1.1.19) formulas. The fundamental theorem about zeros of functions in the Hardy–Hilbert space (Corollary 2.4.10) is the central ingredient of a beautiful proof that every continuous function on $[0,1]$ can be uniformly approximated by polynomials with prime exponents (Corollary 2.5.3). The Hardy–Hilbert space context is necessary to understand the structure of the invariant subspaces of the unilateral shift (Theorem 2.2.12). Conversely, properties of the unilateral shift operator are useful in obtaining results on factorizations of analytic functions (e.g., Theorem 2.3.4) and on other aspects of analytic functions (e.g., Theorem 2.3.3).

The study of Toeplitz operators on the Hardy–Hilbert space is the most natural way of deriving many of the properties of classical Toeplitz matrices (e.g., Theorem 3.3.18), and the study of Hankel operators is the best approach to many results about Hankel matrices (e.g., Theorem 4.3.1). Composition operators are an interesting way of looking at the classical concept of subordination of analytic functions (Corollary 5.1.8). And so on; you'll have to read this entire book (and all the references, and all the references in the references!) to see all the examples that could be listed.

Most of the material discussed in this text was developed by mathematicians whose prime interest was pursuing mathematical beauty. It has turned out, however, as is often the case with pure mathematics, that there are numerous applications of these results, particularly to problems in engineering. Although we do not treat such applications, references are included in the bibliography.

The Hardy–Hilbert space is the set of all analytic functions whose power series have square-summable coefficients (Definition 1.1.1). This Hilbert space of functions analytic on the disk is customarily denoted by $\boldsymbol{H}^2$. There are $\boldsymbol{H}^p$ spaces (called *Hardy spaces*, in honor of G.H. Hardy) for each $p \geq 1$ (and even for $p \in (0,1)$). The only $\boldsymbol{H}^p$ space that is a Hilbert space is $\boldsymbol{H}^2$, the most-studied of the Hardy spaces. We suggest that it should be called the *Hardy–Hilbert space.* There are also other spaces of analytic functions, including the Bergman and Dirichlet spaces. There has been much study of all of these spaces and of various operators on them.

Our goal is to provide an elementary introduction that will be readable by everyone who has understood first courses in complex analysis and in functional analysis. We feel that the best way to do this is to restrict attention to $\boldsymbol{H}^2$ and the operators on it, since that is the easiest setting in which to introduce the essentials of the subject. We have tried to make the exposition as clear, as self-contained, and as instructive as possible, and to make the proofs sufficiently beautiful that they will have a permanent place in mathematics. A reader who masters the material we present will have acquired a firm foundation for the study of all spaces of analytic functions and all operators on such spaces.

This book arose out of lecture notes from graduate courses that were given at the University of Toronto. It should prove suitable as a textbook for courses offered to beginning graduate students, or even to well-prepared advanced undergraduates. We also hope that it will be useful for independent study by students and by mathematicians who wish to learn a new field. Moreover, the exposition should be accessible to students and researchers in those aspects of engineering that rely on this material.

It is our view that a course based on this text would appropriately contribute to the general knowledge of any graduate student in mathematics, whatever field the student might ultimately pursue. In addition, a thorough understanding of this material will provide the necessary background for those who wish to pursue research on related topics. There are a number of excellent books, listed in the references, containing much more extensive treatments of some of the topics covered. There are also references to selected papers of interest. A brief guide to further study is given in the last chapter of this book.

The mathematics presented in this book has its origins in complex analysis, the foundations of which were laid by Cauchy almost 200 years ago. The study of the Hardy–Hilbert space began in the early part of the twentieth century. Hankel operators were first studied toward the end of the nineteenth century, the study of Toeplitz operators was begun early in the twentieth century, and composition operators per se were first investigated in the middle of the twentieth century. There is much current research on properties of Toeplitz, Hankel, composition, and related operators. Thus the material contained in this book was developed by many mathematicians over many decades, and still continues to be the subject of research.

Some references to the development of this subject are given in the "Notes and Remarks" sections at the end of each chapter. We are greatly indebted to the mathematicians cited in these sections and in the references at the end of the book. Moreover, it should be recognized that many other mathematicians have contributed ideas that have become so intrinsic to the subject that their history is difficult to trace.

Our approach to this material has been strongly influenced by the books of Ronald Douglas [16], Peter Duren [17], Paul Halmos [27], and Kenneth Hoffman [32], and by Donald Sarason's lecture notes [49]. The main reason that we have written this book is to provide a gentler introduction to this subject than appears to be available elsewhere.

We are grateful to a number of colleagues for useful comments on preliminary drafts of this book; our special thanks to Sheldon Axler, Paul Bartha, Jaime Cruz-Sampedro, Abie Feintuch, Olivia Gutú, Federico Menéndez-Conde, Eric Nordgren, Steve Power, Heydar Radjavi, Don Sarason, and Nina Zorboska. Moreover, we would like to express our appreciation to Eric Nordgren for pointing out several quite subtle errors in previous drafts. We also thank

Sheldon Axler, Kenneth Ribet, and Mark Spencer for their friendly and encouraging editorial support, and Joel Chan for his TEXnical assistance.

It is very rare that a mathematics book is completely free of errors. We would be grateful if readers who notice mistakes or have constructive criticism would notify us by writing to one of the e-mail addresses given below. We anticipate posting a list of errata on the website

http://www.math.toronto.edu/rosent

Rubén A. Martínez-Avendaño
Centro de Investigación en Matemáticas
Universidad Autónoma del Estado de Hidalgo
Pachuca, Mexico
rubenma@uaeh.edu.mx

Peter Rosenthal
Department of Mathematics
University of Toronto
Toronto, Canada
rosent@math.toronto.edu

August 2006

Contents

Chapter 1

Introduction

In this chapter, we introduce the main definitions and establish some fundamental properties of the Hardy–Hilbert space that we use throughout this book. We also describe various results from functional analysis that are required, including some properties of the spectrum and of invariant subspaces.

1.1 The Hardy–Hilbert Space

The most familiar Hilbert space is called ℓ^2 and consists of the collection of square-summable sequences of complex numbers. That is,

$$\ell^2 = \left\{ \{a_n\}_{n=0}^{\infty} : \sum_{n=0}^{\infty} |a_n|^2 < \infty \right\}.$$

Addition of vectors and multiplication of vectors by complex numbers is performed componentwise. The norm of the vector $\{a_n\}_{n=0}^{\infty}$ is

$$\|\{a_n\}_{n=0}^{\infty}\| = \left(\sum_{n=0}^{\infty} |a_n|^2 \right)^{1/2}$$

and the inner product of the vectors $\{a_n\}_{n=0}^{\infty}$ and $\{b_n\}_{n=0}^{\infty}$ is

$$(\{a_n\}_{n=0}^{\infty}, \{b_n\}_{n=0}^{\infty}) = \sum_{n=0}^{\infty} a_n \overline{b_n}.$$

The space ℓ^2 is separable, and all infinite-dimensional separable complex Hilbert spaces are isomorphic to each other ([12, p. 20], [28, pp. 30–31], [55, p. 90]). Nonetheless, it is often useful to consider particular Hilbert spaces that have additional structure. The space on which we will concentrate, the Hardy–Hilbert space, is a separable Hilbert space whose elements are analytic functions.

Definition 1.1.1. The *Hardy–Hilbert space*, to be denoted $\boldsymbol{H}^2$, consists of all analytic functions having power series representations with square-summable complex coefficients. That is,

$$\boldsymbol{H}^2 = \left\{ f \,:\, f(z) = \sum_{n=0}^{\infty} a_n z^n \text{ and } \sum_{n=0}^{\infty} |a_n|^2 < \infty \right\}.$$

The inner product on $\boldsymbol{H}^2$ is defined by

$$(f, g) = \sum_{n=0}^{\infty} a_n \, \overline{b_n}$$

for

$$f(z) = \sum_{n=0}^{\infty} a_n z^n \quad \text{and} \quad g(z) = \sum_{n=0}^{\infty} b_n z^n.$$

The norm of the vector $f(z) = \sum_{n=0}^{\infty} a_n z^n$ is

$$\|f\| = \left(\sum_{n=0}^{\infty} |a_n|^2 \right)^{1/2}.$$

The mapping $\{a_n\}_{n=0}^{\infty} \longmapsto \sum_{n=0}^{\infty} a_n z^n$ is clearly an isomorphism from ℓ^2 onto $\boldsymbol{H}^2$. Thus, in particular, $\boldsymbol{H}^2$ is a Hilbert space.

Theorem 1.1.2. *Every function in $\boldsymbol{H}^2$ is analytic on the open unit disk.*

Proof. Let $f(z) = \sum_{n=0}^{\infty} a_n z^n$ and $|z_0| < 1$; it must be shown that $\sum_{n=0}^{\infty} a_n z_0^n$ converges. Since $|z_0| < 1$, the geometric series $\sum_{n=0}^{\infty} |z_0|^n$ converges. There exists a K such that $|a_n| \leq K$ for all n (since $\{a_n\}$ is in ℓ^2). Thus $\sum_{n=0}^{\infty} |a_n z_0^n| \leq K \sum_{n=0}^{\infty} |z_0|^n$; hence $\sum_{n=0}^{\infty} a_n z_0^n$ converges absolutely. □

Notation 1.1.3. The *open unit disk* in the complex plane, $\{z \in \mathbb{C} : |z| < 1\}$, will be denoted by $\mathbb{D}$, and the *unit circle*, $\{z \in \mathbb{C} : |z| = 1\}$, will be denoted by S^1.

The space $\boldsymbol{H}^2$ obviously contains all polynomials and many other analytic functions.

Example 1.1.4. *For each point $e^{i\theta_0} \in S^1$, there is a function in $\boldsymbol{H}^2$ that is not analytic at $e^{i\theta_0}$.*

Proof. Define f_{θ_0} by

$$f_{\theta_0}(z) = \sum_{n=1}^{\infty} \frac{e^{-in\theta_0}}{n} z^n \qquad \text{for } z \in \mathbb{D}.$$

Since $\left\{ \frac{e^{-in\theta_0}}{n} \right\} \in \ell^2$, $f_{\theta_0} \in \boldsymbol{H}^2$. As z approaches $e^{i\theta_0}$ from within $\mathbb{D}$, $|f_{\theta_0}(z)|$ approaches infinity. Hence there is no way of defining f_{θ_0} so that it is analytic at $e^{i\theta_0}$. □

This example can be strengthened: there are functions in $\boldsymbol{H}^2$ that are not analytic at any point in S^1 (see Example 2.4.15 below).

It is easy to find examples of functions analytic on $\mathbb{D}$ that are not in $\boldsymbol{H}^2$.

Example 1.1.5. *The function $f(z) = \frac{1}{1-z}$ is analytic on $\mathbb{D}$ but is not in $\boldsymbol{H}^2$.*

Proof. Since $\frac{1}{1-z} = \sum_{n=0}^{\infty} z^n$, the coefficients of f are not square-summable. □

Bounded linear functionals (i.e., continuous linear mappings from a linear space into the space of complex numbers) are very important in the study of linear operators. The "point evaluations" are particularly useful linear functionals on $\boldsymbol{H}^2$.

Theorem 1.1.6. *For every $z_0 \in \mathbb{D}$, the mapping $f \longmapsto f(z_0)$ is a bounded linear functional on $\boldsymbol{H}^2$.*

Proof. Fix $z_0 \in \mathbb{D}$. Note that the Cauchy–Schwarz inequality yields

$$\begin{aligned} |f(z_0)| &= \left| \sum_{n=0}^{\infty} a_n z_0^n \right| \\ &\leq \left(\sum_{n=0}^{\infty} |a_n|^2 \right)^{1/2} \left(\sum_{n=0}^{\infty} |z_0|^{2n} \right)^{1/2} \\ &= \left(\sum_{n=0}^{\infty} |z_0|^{2n} \right)^{1/2} \|f\|. \end{aligned}$$

It is obvious that evaluation at z_0 is a linear mapping of $\boldsymbol{H}^2$ into $\mathbb{C}$. Thus the mapping is a bounded linear functional of norm at most $\left(\sum_{n=0}^{\infty} |z_0|^{2n}\right)^{1/2}$. □

The Riesz representation theorem states that every linear functional on a Hilbert space can be represented by an inner product with a vector in the space ([12, p. 13], [28, pp. 31–32], [55, p. 142]). This representation can be explicitly stated for point evaluations on $\boldsymbol{H}^2$.

Definition 1.1.7. For $z_0 \in \mathbb{D}$, the function k_{z_0} defined by

$$k_{z_0}(z) = \sum_{n=0}^{\infty} \overline{z_0}^n z^n = \frac{1}{1 - \overline{z_0}\, z}$$

is called the *reproducing kernel* for z_0 in $\boldsymbol{H}^2$.

It is obvious that $k_{z_0} \in \boldsymbol{H}^2$. Point evaluations are representable as inner products with reproducing kernels.

Theorem 1.1.8. *For $z_0 \in \mathbb{D}$ and $f \in \boldsymbol{H}^2$, $f(z_0) = (f, k_{z_0})$ and $\|k_{z_0}\| = \left(1 - |z_0|^2\right)^{-1/2}$.*

Proof. Writing k_{z_0} as $\sum_{n=0}^{\infty} \overline{z_0}^n z^n$ yields

$$(f, k_{z_0}) = \sum_{n=0}^{\infty} a_n z_0^n = f(z_0),$$

and

$$\|k_{z_0}\|^2 = \sum_{n=0}^{\infty} |\overline{z_0}|^{2n}.$$

Since $\displaystyle\sum_{n=0}^{\infty} |\overline{z_0}|^{2n} = \frac{1}{1 - |z_0|^2}$, it follows that $\displaystyle\|k_{z_0}\| = \frac{1}{(1 - |z_0|^2)^{1/2}}$. □

Our first application of reproducing kernels is in establishing the following relationship between convergence in $\boldsymbol{H}^2$ and convergence as analytic functions.

Theorem 1.1.9. *If $\{f_n\} \to f$ in $\boldsymbol{H}^2$, then $\{f_n\} \to f$ uniformly on compact subsets of $\mathbb{D}$.*

Proof. For a fixed $z_0 \in \mathbb{D}$, we have

$$|f_n(z_0) - f(z_0)| = |(f_n - f, k_{z_0})| \leq \|f_n - f\|\ \|k_{z_0}\|.$$

If K is a compact subset of $\mathbb{D}$, then there exists an M such that $\|k_{z_0}\| \leq M$ for all $z_0 \in K$ (M can be taken to be the supremum of $\frac{1}{\sqrt{1-|z_0|^2}}$ for $z_0 \in K$). Hence

$$|f_n(z_0) - f(z_0)| \leq M\|f_n - f\| \qquad \text{for all } z_0 \in K,$$

which clearly implies the theorem. □

Thus convergence in the Hilbert space norm implies convergence in the standard topology on the space of all analytic functions on $\mathbb{D}$.

The Hardy–Hilbert space can also be viewed as a subspace of another well-known Hilbert space.

We denote by $\boldsymbol{L}^2 = \boldsymbol{L}^2(S^1)$ the Hilbert space of square-integrable functions on S^1 with respect to Lebesgue measure, normalized so that the measure of the entire circle is 1. The inner product is given by

$$(f, g) = \frac{1}{2\pi} \int_0^{2\pi} f(e^{i\theta}) \, \overline{g(e^{i\theta})} \, d\theta,$$

where $d\theta$ denotes the ordinary (not normalized) Lebesgue measure on $[0, 2\pi]$. Therefore the norm of the function f in $\boldsymbol{L}^2$ is given by

$$\|f\| = \left(\frac{1}{2\pi} \int_0^{2\pi} |f(e^{i\theta})|^2 \, d\theta \right)^{1/2}.$$

We use the same symbols to denote the norms and inner products of all the Hilbert spaces we consider. It should be clear from the context which norm or inner product is being used.

As is customary, we often abuse the language and view $\boldsymbol{L}^2$ as a space of functions rather than as a space of equivalence classes of functions. We then say that two $\boldsymbol{L}^2$ functions are equal when we mean they are equal almost everywhere with respect to normalized Lebesgue measure. We will sometimes omit the words "almost everywhere" (or "a.e.") unless we wish to stress that equality holds only in that sense.

For each integer n, let $e_n(e^{i\theta}) = e^{in\theta}$, regarded as a function on S^1. It is well known that the set $\{e_n : n \in \mathbb{Z}\}$ forms an orthonormal basis for $\boldsymbol{L}^2$ ([2, p. 24], [12, p. 21], [42, p. 48], [47, pp. 89–92]). We define the space $\widetilde{\boldsymbol{H}}^2$ as the following subspace of $\boldsymbol{L}^2$:

$$\widetilde{\boldsymbol{H}}^2 = \{\widetilde{f} \in \boldsymbol{L}^2 : (\widetilde{f}, e_n) = 0 \text{ for } n < 0\}.$$

That is, $\widetilde{f} \in \widetilde{\boldsymbol{H}}^2$ if its Fourier series is of the form

$$\widetilde{f}(e^{i\theta}) = \sum_{n=0}^{\infty} a_n e^{in\theta} \quad \text{with} \quad \sum_{n=0}^{\infty} |a_n|^2 < \infty.$$

It is clear that $\widetilde{\boldsymbol{H}}^2$ is a closed subspace of $\boldsymbol{L}^2$. Also, there is a natural identification between $\widetilde{\boldsymbol{H}}^2$ and $\boldsymbol{H}^2$. Namely, we identify the function $\widetilde{f} \in \widetilde{\boldsymbol{H}}^2$ having Fourier series $\sum_{n=0}^{\infty} a_n e^{in\theta}$ with the analytic function $f(z) = \sum_{n=0}^{\infty} a_n z^n$.

This identification is clearly an isomorphism between $\boldsymbol{H}^2$ and $\widetilde{\boldsymbol{H}}^2$. Of course, this identification, although natural, does not describe (at least in an obvious way) the relationship between $f \in \boldsymbol{H}^2$ and $\widetilde{f} \in \widetilde{\boldsymbol{H}}^2$ as *functions*. We proceed to investigate this.

Let $\widetilde{f} \in \widetilde{\boldsymbol{H}}^2$ have Fourier series $\sum_{n=0}^{\infty} a_n e^{in\theta}$ and $f \in \boldsymbol{H}^2$ have power series $f(z) = \sum_{n=0}^{\infty} a_n z^n$. For $0 < r < 1$, let f_r be defined by

$$f_r(e^{i\theta}) = f(re^{i\theta}) = \sum_{n=0}^{\infty} a_n r^n e^{in\theta}.$$

Clearly, $f_r \in \widetilde{\boldsymbol{H}}^2$ for every such r.

Theorem 1.1.10. *Let $\widetilde{f}$ and f_r be defined as above. Then*

$$\lim_{r \to 1^-} \|\widetilde{f} - f_r\| = 0 \quad \textit{in } \widetilde{\boldsymbol{H}}^2.$$

Proof. Let $\varepsilon > 0$ be given. Since $\sum_{n=0}^{\infty} |a_n|^2 < \infty$, we can choose a natural number n_0 such that

$$\sum_{n=n_0}^{\infty} |a_n|^2 < \frac{\varepsilon}{2}.$$

Now choose s between 0 and 1 such that for every $r \in (s, 1)$ we have

$$\sum_{n=0}^{n_0-1} |a_n|^2 (1 - r^n)^2 < \frac{\varepsilon}{2}.$$

Then, since

$$\|\widetilde{f} - f_r\|^2 = \left\| \sum_{n=0}^{\infty} (a_n - a_n r^n) e^{in\theta} \right\|^2 = \sum_{n=0}^{\infty} |a_n|^2 (1 - r^n)^2,$$

it follows that

$$\begin{aligned}
\|\widetilde{f} - f_r\|^2 &= \sum_{n=0}^{n_0-1} |a_n|^2 (1 - r^n)^2 + \sum_{n=n_0}^{\infty} |a_n|^2 (1 - r^n)^2 \\
&< \frac{\varepsilon}{2} + \sum_{n=n_0}^{\infty} |a_n|^2 \\
&< \frac{\varepsilon}{2} + \frac{\varepsilon}{2} \\
&= \varepsilon.
\end{aligned}$$

□

An important consequence is the following.

Corollary 1.1.11. *For each f in $\boldsymbol{H}^2$, there exists an increasing sequence $\{r_n\}$ of positive numbers converging to 1 such that*

$$\lim_{n\to\infty} f\left(r_n e^{i\theta}\right) = \widetilde{f}\left(e^{i\theta}\right)$$

for almost all θ.

Proof. It is well known that convergence in $\boldsymbol{L}^2$ implies that a subsequence converges pointwise almost everywhere [47, p. 68], so this follows from the previous theorem. □

We prove a stronger result at the end of this section: $\lim\limits_{r\to 1^-} f(re^{i\theta}) = \widetilde{f}(e^{i\theta})$ for almost all θ (that is, not just for a subsequence $\{r_n\} \to 1$).

There is an alternative definition of the Hardy–Hilbert space.

Theorem 1.1.12. *Let f be analytic on $\mathbb{D}$. Then $f \in \boldsymbol{H}^2$ if and only if*

$$\sup_{0<r<1} \frac{1}{2\pi}\int_0^{2\pi} \left|f(re^{i\theta})\right|^2 \, d\theta < \infty.$$

Moreover, for $f \in \boldsymbol{H}^2$,

$$\|f\|^2 = \sup_{0<r<1} \frac{1}{2\pi}\int_0^{2\pi} \left|f(re^{i\theta})\right|^2 \, d\theta.$$

Proof. Let f be an analytic function on $\mathbb{D}$ with power series

$$f(z) = \sum_{n=0}^{\infty} a_n z^n.$$

Then, for $0 < r < 1$,

$$|f(re^{i\theta})|^2 = \sum_{n=0}^{\infty}\sum_{m=0}^{\infty} a_n \overline{a_m} r^{n+m} e^{i(n-m)\theta}.$$

Since

$$\frac{1}{2\pi}\int_0^{2\pi} e^{i(n-m)\theta} d\theta = \delta_{n,m},$$

integrating the expression above for $|f(re^{i\theta})|^2$ and dividing by 2π results in

$$\frac{1}{2\pi}\int_0^{2\pi} \left|f(re^{i\theta})\right|^2 \, d\theta = \sum_{n=0}^{\infty} |a_n|^2 r^{2n}.$$

If $f \in \boldsymbol{H}^2$, then $\sum_{n=0}^{\infty} |a_n|^2 r^{2n} \leq \|f\|^2$ for every r in $[0,1)$. Thus

$$\sup_{0<r<1} \frac{1}{2\pi} \int_0^{2\pi} \left|f(re^{i\theta})\right|^2 \, d\theta \leq \|f\|^2 < \infty.$$

Conversely, assume that the above supremum is finite. As shown above,

$$\frac{1}{2\pi} \int_0^{2\pi} \left|f(re^{i\theta})\right|^2 \, d\theta = \sum_{n=0}^{\infty} |a_n|^2 r^{2n}.$$

If $f \notin \boldsymbol{H}^2$, the right-hand side can be made arbitrarily large by taking r close to 1. This would contradict the assumption that the supremum of the left side of the equation is finite.

Note that the above also shows that, for $f \in \boldsymbol{H}^2$,

$$\|f\|^2 = \sup_{0<r<1} \frac{1}{2\pi} \int_0^{2\pi} \left|f(re^{i\theta})\right|^2 \, d\theta.$$

□

Corollary 1.1.13. *For any function f analytic on the disk, the function*

$$M(r) = \frac{1}{2\pi} \int_0^{2\pi} |f(re^{i\theta})|^2 \, d\theta$$

is increasing. Therefore $\lim_{r \to 1^-} M(r) = \sup_{0<r<1} M(r)$, and hence the function f is in $\boldsymbol{H}^2$ if and only if $\lim_{r \to 1^-} M(r) < \infty$, in which case $\lim_{r \to 1^-} M(r) = \|f\|^2$.

Proof. This follows immediately from the formula

$$\frac{1}{2\pi} \int_0^{2\pi} \left|f(re^{i\theta})\right|^2 \, d\theta = \sum_{n=0}^{\infty} |a_n|^2 r^{2n}$$

established in the course of the proof of the preceding theorem. □

The next example will be useful in computing eigenvectors of hyperbolic composition operators (see Theorem 5.4.10 in Chapter 5).

Example 1.1.14. *For $s \in (0, \frac{1}{2})$, the function*

$$\frac{1}{(1-z)^s}$$

is in $\boldsymbol{H}^2$. (Recall that $(1-z)^s = \exp(s \log(1-z))$, where log *denotes the principal branch of the logarithm.)*

Proof. Fix any s in $(0, \frac{1}{2})$ and let

$$f(z) = \frac{1}{(1-z)^s}.$$

For each r, let

$$M(r) = \frac{1}{2\pi} \int_0^{2\pi} |f(re^{i\theta})|^2 d\theta.$$

By Theorem 1.1.12, it suffices to show that there exists an M such that $M(r) \leq M$ for all $r \in (0,1)$.

Fix an r. An easy computation gives

$$|1 - re^{i\theta}|^2 = 1 + r^2 - 2r\cos\theta,$$

so

$$\left|\frac{1}{(1-re^{i\theta})^s}\right|^2 = \frac{1}{(1+r^2-2r\cos\theta)^s}.$$

To estimate $M(r)$, first note that the periodicity of cosine implies that

$$\begin{aligned}\frac{1}{2\pi}\int_0^{2\pi} \frac{d\theta}{(1+r^2-2r\cos\theta)^s} &= \frac{1}{\pi}\int_0^{\pi} \frac{d\theta}{(1+r^2-2r\cos\theta)^s} \\ &= \frac{1}{\pi}\int_0^{\frac{\pi}{2}} \frac{d\theta}{(1+r^2-2r\cos\theta)^s} + \frac{1}{\pi}\int_{\frac{\pi}{2}}^{\pi} \frac{d\theta}{(1+r^2-2r\cos\theta)^s}.\end{aligned}$$

We separately estimate each of these integrals. To estimate the first integral, begin by noting that $1 + r^2 - 2r\cos\theta = (r - \cos\theta)^2 + \sin^2\theta$, which is greater than or equal to $\sin^2\theta$. Hence

$$\frac{1}{\pi}\int_0^{\frac{\pi}{2}} \frac{d\theta}{(1+r^2-2r\cos\theta)^s} \leq \frac{1}{\pi}\int_0^{\frac{\pi}{2}} \frac{d\theta}{\sin^{2s}\theta}.$$

To see that this latter integral converges, write

$$\frac{1}{\pi}\int_0^{\frac{\pi}{2}} \frac{d\theta}{\sin^{2s}\theta} = \frac{1}{\pi}\int_0^{\frac{\pi}{4}} \frac{d\theta}{\sin^{2s}\theta} + \frac{1}{\pi}\int_{\frac{\pi}{4}}^{\frac{\pi}{2}} \frac{d\theta}{\sin^{2s}\theta}.$$

It suffices to show that

$$\frac{1}{\pi}\int_0^{\frac{\pi}{4}} \frac{d\theta}{\sin^{2s}\theta}$$

converges.

It is easily verified that $\theta \leq \tan\theta$ for $\theta \in [0, \frac{\pi}{2})$. Hence $\sin\theta \geq \theta\cos\theta$, so, for $\theta \in [0, \frac{\pi}{4})$, $\sin\theta \geq \frac{1}{\sqrt{2}}\,\theta$. Therefore

$$\frac{1}{\pi}\int_0^{\frac{\pi}{4}} \frac{d\theta}{\sin^{2s}\theta} \leq \frac{2^s}{\pi}\int_0^{\frac{\pi}{4}} \frac{d\theta}{\theta^{2s}}.$$

As is well known and easily verified,

$$\int_0^{\frac{\pi}{4}} \frac{d\theta}{\theta^{2s}}$$

converges when $2s < 1$, and thus when $s < \frac{1}{2}$.

We must now estimate

$$\frac{1}{\pi} \int_{\frac{\pi}{2}}^{\pi} \frac{d\theta}{(1 + r^2 - 2r\cos\theta)^s}.$$

Since $\cos\theta \leq 0$ for $\theta \in [\frac{\pi}{2}, \pi]$,

$$1 + r^2 - 2r\cos\theta \geq 1 + r^2$$

for such θ. Therefore

$$\frac{1}{\pi} \int_{\frac{\pi}{2}}^{\pi} \frac{d\theta}{(1 + r^2 - 2r\cos\theta)^s} \leq \frac{1}{\pi} \int_{\frac{\pi}{2}}^{\pi} \frac{d\theta}{(1 + r^2)^s} \leq \frac{1}{2(1 + r^2)^s} \leq \frac{1}{2}.$$

Thus, for every $r \in [0, 1)$,

$$M(r) \leq \frac{1}{\pi} \int_{\frac{\pi}{4}}^{\frac{\pi}{2}} \frac{d\theta}{\sin^{2s}\theta} + \frac{2^s}{\pi} \int_0^{\frac{\pi}{4}} \frac{d\theta}{\theta^{2s}} + \frac{1}{2}.$$

Since this bound on $M(r)$ is independent of r, it follows from Theorem 1.1.12 that f is in $\boldsymbol{H}^2$. □

Another space of analytic functions arises in the study of operators on $\boldsymbol{H}^2$.

Definition 1.1.15. The space $\boldsymbol{H}^\infty$ consists of all the functions that are analytic and bounded on the open unit disk. The vector operations are the usual pointwise addition of functions and multiplication by complex scalars. The norm of a function f in $\boldsymbol{H}^\infty$ is defined by $\|f\|_\infty = \sup\{|f(z)| : z \in \mathbb{D}\}$.

Since convergence in the norm on $\boldsymbol{H}^\infty$ implies uniform convergence on the disk, it is easily seen that $\boldsymbol{H}^\infty$ is a Banach space.

Corollary 1.1.16. *Every function in* $\boldsymbol{H}^\infty$ *is in* $\boldsymbol{H}^2$.

Proof. This follows immediately from the characterization of $\boldsymbol{H}^2$ given in Theorem 1.1.12. □

We shall see that multiplication by a function in $\boldsymbol{H}^\infty$ induces a bounded linear operator on $\boldsymbol{H}^2$. Such operators, called analytic Toeplitz operators, play an important role in the sequel (see Chapter 3).

Theorem 1.1.17. *If $f \in \boldsymbol{H}^\infty$ and f is not a constant, then $|f(z)| < \|f\|_\infty$ for all $z \in \mathbb{D}$.*

Proof. This is an immediate consequence of the maximum modulus theorem ([9, pp. 79, 128], [47, p. 212]). □

The following interesting collection of functions in $\boldsymbol{H}^\infty$ will be used, in combination with the functions of Example 1.1.14, in describing eigenvectors of hyperbolic composition operators (see Theorem 5.4.10 in Chapter 5).

Example 1.1.18. *For each real number t, the function*

$$\left(\frac{1+z}{1-z}\right)^{it}$$

is in $\boldsymbol{H}^\infty$. (Recall that $w^{it} = \exp(it \log w)$, where $\log$ is the principal branch of the logarithm.)

Proof. Note that, for every $z \in \mathbb{D}$, the number

$$w = \frac{1+z}{1-z}$$

is in the open right half-plane. For each such w,

$$w^{it} = \exp(it \log w) = \exp(it(\log r + i\theta)),$$

where $w = re^{i\theta}$ and θ is in $(-\frac{\pi}{2}, \frac{\pi}{2})$. It follows that $|w^{it}| = \exp(-t\theta)$, which is at most $\exp\left(\frac{|t|\pi}{2}\right)$. Hence

$$\left|\left(\frac{1+z}{1-z}\right)^{it}\right| \leq \exp\left(\frac{|t|\pi}{2}\right)$$

for all $z \in \mathbb{D}$. □

Reproducing kernels can be used to give a proof of a special case of the Cauchy integral formula.

Theorem 1.1.19 (Cauchy Integral Formula). *If f is analytic on an open set containing $\overline{\mathbb{D}}$ and $z_0 \in \mathbb{D}$, then*

$$f(z_0) = \frac{1}{2\pi i} \int_{S^1} \frac{f(z)}{z - z_0}\, dz.$$

Proof. Since f is analytic on $\overline{\mathbb{D}}$, Corollary 1.1.11 implies that $\widetilde{f}(e^{i\theta}) = f(e^{i\theta})$ for all θ. Note that k_{z_0} is continuous on $\overline{\mathbb{D}}$, and therefore

$$\widetilde{k_{z_0}}(e^{i\theta}) = \frac{1}{1 - \overline{z_0} e^{i\theta}}.$$

For $z_0 \in \mathbb{D}$,

$$\begin{aligned} f(z_0) = (f, k_{z_0}) = \left(\widetilde{f}, \widetilde{k_{z_0}}\right) &= \frac{1}{2\pi} \int_0^{2\pi} \widetilde{f}(e^{i\theta}) \overline{\widetilde{k_{z_0}}(e^{i\theta})} \, d\theta \\ &= \frac{1}{2\pi} \int_0^{2\pi} \widetilde{f}(e^{i\theta}) \frac{1}{1 - z_0 \, e^{-i\theta}} \, d\theta \\ &= \frac{1}{2\pi i} \int_0^{2\pi} \frac{\widetilde{f}(e^{i\theta})}{e^{i\theta} - z_0} \, i e^{i\theta} \, d\theta. \end{aligned}$$

Letting $z = e^{i\theta}$, this expression becomes

$$\frac{1}{2\pi i} \int_{S^1} \frac{\widetilde{f}(z)}{z - z_0} \, dz.$$

Thus

$$f(z_0) = \frac{1}{2\pi i} \int_{S^1} \frac{\widetilde{f}(z)}{z - z_0} \, dz.$$

Since $f(z) = \widetilde{f}(z)$ when $|z| = 1$, we have

$$f(z_0) = \frac{1}{2\pi i} \int_{S^1} \frac{f(z)}{z - z_0} \, dz.$$

□

A similar approach can be taken to the Poisson integral formula.

Definition 1.1.20. For $0 \leq r < 1$ and $\psi \in [0, 2\pi]$, the *Poisson kernel* is defined by

$$P_r(\psi) = \frac{1 - r^2}{1 - 2r \cos \psi + r^2}.$$

Observe that $P_r(\psi) > 0$ for all $r \in [0, 1)$ and all ψ, since

$$1 - r^2 > 0 \quad \text{and} \quad 1 - 2r \cos \psi + r^2 \geq (1 - r)^2 > 0.$$

Theorem 1.1.21 (Poisson Integral Formula). *If f is in $\boldsymbol{H}^2$ and re^{it} is in $\mathbb{D}$, then*

$$f(re^{it}) = \frac{1}{2\pi} \int_0^{2\pi} \widetilde{f}(e^{i\theta}) P_r(\theta - t) \, d\theta.$$

Proof. Let $z_0 \in \mathbb{D}$. Since

$$\widetilde{k_{z_0}}(e^{i\theta}) = \frac{1}{1 - \overline{z_0}e^{i\theta}},$$

we have

$$f(z_0) = (f, k_{z_0}) = \left(\widetilde{f}, \widetilde{k_{z_0}}\right) = \frac{1}{2\pi}\int_0^{2\pi} \frac{\widetilde{f}(e^{i\theta})}{1 - z_0 e^{-i\theta}}\, d\theta.$$

But

$$\frac{1}{1 - z_0 e^{-i\theta}} = 1 + z_0 e^{-i\theta} + z_0^2 e^{-2i\theta} + z_0^3 e^{-3i\theta} + \cdots,$$

so the function

$$\frac{1}{1 - z_0 e^{-i\theta}} - 1$$

has all its Fourier coefficients corresponding to negative indices equal to 0. It is therefore orthogonal to $\widetilde{f}$, so

$$\frac{1}{2\pi}\int_0^{2\pi} \widetilde{f}(e^{i\theta}) \overline{\left(\frac{1}{1 - z_0 e^{-i\theta}} - 1\right)}\, d\theta = 0.$$

Adding this integral to the one displayed above for $f(z_0)$ yields

$$f(z_0) = \frac{1}{2\pi}\int_0^{2\pi} \widetilde{f}(e^{i\theta}) \left(\frac{1}{1 - z_0 e^{-i\theta}} + \frac{1}{1 - \overline{z_0}e^{i\theta}} - 1\right) d\theta.$$

If $z_0 = re^{it}$, a very straightforward calculation shows that

$$\frac{1}{1 - z_0 e^{-i\theta}} + \frac{1}{1 - \overline{z_0}e^{i\theta}} - 1 = \frac{1 - r^2}{1 - 2r\cos(\theta - t) + r^2}.$$

But

$$P_r(\theta - t) = \frac{1 - r^2}{1 - 2r\cos(\theta - t) + r^2},$$

so

$$f(re^{it}) = \frac{1}{2\pi}\int_0^{2\pi} \widetilde{f}(e^{i\theta}) P_r(\theta - t)\, d\theta.$$

□

The following fact will be needed in subsequent applications of the above theorem.

Corollary 1.1.22. *For $r \in [0, 1)$ and t any real number,*

$$\frac{1}{2\pi}\int_0^{2\pi} P_r(\theta - t)\, d\theta = 1.$$

Proof. This is an immediate application of Theorem 1.1.21 to the case where f is the constant function 1. □

Definition 1.1.23. The measurable function ϕ on S^1 is *essentially bounded* if there exists some M_0 such that the measure of

$$\{e^{i\theta} : |\phi(e^{i\theta})| > M_0\}$$

is 0. The space $\boldsymbol{L}^\infty$ is the collection of all (equivalence classes modulo sets of measure zero of) essentially bounded measurable functions. The *essential norm* of the function $\phi \in \boldsymbol{L}^\infty$, denoted $\|\phi\|_\infty$, is defined by

$$\|\phi\|_\infty = \inf\left\{M : \text{ the measure of } \{e^{i\theta} : |\phi(e^{i\theta})| > M\} \text{ is } 0\right\}.$$

Observe that, for $\phi \in \boldsymbol{L}^\infty$, the inequality $|\phi(e^{i\theta})| \leq \|\phi\|_\infty$ holds for almost all θ.

Corollary 1.1.24. *Let $f \in \boldsymbol{H}^2$ and suppose that $|\widetilde{f}(e^{i\theta})| \leq K$ a.e. Then $|f(z)| \leq K$ for all $z \in \mathbb{D}$. In particular, a function in $\boldsymbol{H}^2$ whose boundary function is in $\boldsymbol{L}^\infty$ must be in $\boldsymbol{H}^\infty$.*

Proof. Recall that $P_r(\theta) > 0$ for all θ and $0 \leq r < 1$. For $re^{it} \in \mathbb{D}$, applying the Poisson integral formula (Theorem 1.1.21) to f yields

$$\begin{aligned}
|f(re^{it})| &= \left|\frac{1}{2\pi}\int_0^{2\pi} \widetilde{f}(e^{i\theta})P_r(\theta - t)\, d\theta\right| \\
&\leq \frac{1}{2\pi}\int_0^{2\pi} |\widetilde{f}(e^{i\theta})|P_r(\theta - t)\, d\theta \\
&\leq K\frac{1}{2\pi}\int_0^{2\pi} P_r(\theta - t)\, d\theta \\
&= K,
\end{aligned}$$

by the previous corollary. Therefore $|f(z)| \leq K$ for all $z \in \mathbb{D}$, as desired. □

To further clarify the relation between f and $\widetilde{f}$ requires a theorem known as Fatou's theorem, which we prove below.

First, recall the following definition.

Definition 1.1.25. Let α be a complex-valued function of a real variable. The *symmetric derivative* of α at t is defined to be

$$\lim_{h \to 0} \frac{\alpha(t + h) - \alpha(t - h)}{2h},$$

if the limit exists.

Clearly, if α is differentiable at t the symmetric derivative will exist and will equal $\alpha'(t)$, since

$$\frac{\alpha(t+h)-\alpha(t)}{h}+\frac{\alpha(t-h)-\alpha(t)}{-h}=2\frac{\alpha(t+h)-\alpha(t-h)}{2h}.$$

The converse is not true in general (see Exercise 1.9).

We require a generalization of the Riemann integral known as the Riemann–Stieltjes integral, defined as follows. Let α be a complex-valued function having bounded variation on the interval $[a,b]$. (This is equivalent to the real and imaginary parts of α having bounded variation, and a real-valued function has bounded variation if and only if it is the difference of two nondecreasing functions on the interval.) For f any continuous function on $[a,b]$, the Riemann–Stieltjes integral of f with respect to α, denoted by $\int_a^b f(t)\,d\alpha(t)$, is defined to be the limit of Riemann–Stieltjes sums of the form $\sum_{i=0}^{n-1} f(t_i)(\alpha(x_{i+1})-\alpha(x_i))$, where each t_i is in $[x_i,x_{i+1}]$, as the mesh of the partition $\{x_0,x_1,x_2,\ldots,x_n\}$ goes to 0. The proof of the existence of the Riemann–Stieltjes integral of a continuous function with respect to a nondecreasing function α is essentially the same as the proof of the existence of the ordinary Riemann integral, and the extension to general α of bounded variation follows by linearity. (See Apostol [3] for an excellent discussion of Riemann–Stieltjes integrals.)

The basic relationship between a function in $\boldsymbol{H}^2$ and its boundary values will be obtained as a consequence of the following theorem.

Theorem 1.1.26 (Fatou's Theorem). *Let α be a complex-valued function of bounded variation on $[0,2\pi]$ and let u be the function defined on the open unit disk by*

$$u(re^{it})=\frac{1}{2\pi}\int_0^{2\pi}P_r(\theta-t)\,d\alpha(\theta).$$

If the symmetric derivative of α exists at $t_0\in(0,2\pi)$, then

$$\lim_{r\to 1^-}u(re^{it_0})$$

exists and equals the symmetric derivative of α at t_0.

Proof. We need to extend α to the entire real line. Define α on the interval $(k2\pi,(k+1)2\pi]$ for each postive integer k, and on the interval $[k2\pi,(k+1)2\pi)$ for each negative integer k, as

$$\alpha(\theta)=\alpha(\theta-k2\pi)+k(\alpha(2\pi)-\alpha(0)).$$

With this extension of α it can be readily verified that, for each integer k, interval $[a, b]$, and continuous function f of period 2π,

$$\int_a^b f(\theta)\, d\alpha(\theta) = \int_{a+k2\pi}^{b+k2\pi} f(\theta)\, d\alpha(\theta).$$

To proceed with the proof, let L be the symmetric derivative of α at t_0 and let $\varepsilon > 0$. We shall show that there exists $s \in (0,1)$ such that

$$|u(re^{it_0}) - L| < \varepsilon \qquad \text{for all } r \in [s, 1).$$

Since the symmetric derivative of α exists at t_0, we can choose $\delta > 0$ (but also keep $\delta < \pi$) such that

$$\left| \frac{\alpha(t_0 + h) - \alpha(t_0 - h)}{2h} - L \right| < \frac{\varepsilon}{8} \qquad \text{if } 0 < |h| < \delta.$$

Recall that

$$\frac{1}{2\pi} \int_0^{2\pi} P_r(\theta - t)\, d\theta = 1.$$

Thus

$$u(re^{it_0}) - L = \frac{1}{2\pi} \int_0^{2\pi} P_r(\theta - t_0)\, (d\alpha(\theta) - L\, d\theta).$$

Using the property of the extended function α and the fact that P_r is a periodic function with period 2π, the last expression becomes, after a change of variables,

$$u(re^{it_0}) - L = \frac{1}{2\pi} \int_0^{2\pi} P_r(\psi)\, (d\alpha(\psi + t_0) - L\, d\psi).$$

We will separate this integral into several parts. Using the positive number δ obtained above and the triangle inequality, we get

$$\begin{aligned} |u(re^{it_0}) - L| \le & \left| \frac{1}{2\pi} \int_0^{\delta} P_r(\psi)\, (d\alpha(\psi + t_0) - L\, d\psi) \right. \\ & \left. + \frac{1}{2\pi} \int_{2\pi-\delta}^{2\pi} P_r(\psi)\, (d\alpha(\psi + t_0) - L\, d\psi) \right| \\ & + \left| \frac{1}{2\pi} \int_{\delta}^{2\pi-\delta} P_r(\psi)\, (d\alpha(\psi + t_0) - L\, d\psi) \right|. \end{aligned}$$

We first prove that the last term can be made small by taking r sufficiently close to 1. For all $\psi \in [\delta, 2\pi - \delta]$, it is clear that

$$1 - 2r\cos\psi + r^2 \geq 1 - 2r\cos\delta + r^2.$$

Since

$$\lim_{r\to 1^-} (1 - 2r\cos\delta + r^2) = 2 - 2\cos\delta > 0,$$

there exists a positive real number η and a number $s_1 \in (0,1)$ such that

$$1 - 2r\cos\psi + r^2 \geq 1 - 2r\cos\delta + r^2 \geq \eta > 0$$

for all $\psi \in [\delta, 2\pi - \delta]$ and $r \in [s_1, 1)$. Hence

$$P_r(\psi) \leq \frac{1}{\eta}(1 - r^2) \qquad \text{for } \psi \in [\delta, 2\pi - \delta] \text{ and } r \in [s_1, 1).$$

Therefore,

$$\left| \frac{1}{2\pi} \int_\delta^{2\pi-\delta} P_r(\psi)\ (d\alpha(\psi + t_0) - L\ d\psi) \right| \leq \frac{1 - r^2}{2\pi\eta} \int_\delta^{2\pi-\delta} |d\alpha(\psi + t_0) - L\ d\psi|\,.$$

Since $(1 - r^2)$ goes to 0 as $r \to 1^-$, we can choose $s_2 \in [s_1, 1)$ such that

$$\frac{1 - r^2}{2\pi\eta} \int_\delta^{2\pi-\delta} |d\alpha(\psi + t_0) - L\ d\psi| < \frac{\varepsilon}{2}$$

for all $r \in [s_2, 1)$. Hence

$$\left| \frac{1}{2\pi} \int_\delta^{2\pi-\delta} P_r(\psi)\ (d\alpha(\psi + t_0) - L\ d\psi) \right| < \frac{\varepsilon}{2} \qquad \text{for all } r \in [s_2, 1).$$

By periodicity, we obtain

$$\frac{1}{2\pi} \int_0^\delta P_r(\psi)\ (d\alpha(\psi + t_0) - L\ d\psi) + \frac{1}{2\pi} \int_{2\pi-\delta}^{2\pi} P_r(\psi)\ (d\alpha(\psi + t_0) - L\ d\psi)$$
$$= \frac{1}{2\pi} \int_{-\delta}^{\delta} P_r(\psi)\ (d\alpha(\psi + t_0) - L\ d\psi).$$

Integration by parts of this last integral results in

$$\frac{1}{2\pi} \int_{-\delta}^{\delta} P_r(\psi)\ (d\alpha(\psi + t_0) - L\ d\psi)$$
$$= \frac{1}{2\pi} \left(P_r(\delta)(\alpha(\delta + t_0) - L\delta) - P_r(-\delta)(\alpha(-\delta + t_0) + L\delta)\right)$$
$$- \frac{1}{2\pi} \int_{-\delta}^{\delta} (\alpha(\psi + t_0) - L\psi)\, \frac{dP_r}{d\psi}\ d\psi.$$

Observe that, since $P_r(\delta) = P_r(-\delta)$ and, as shown above, $P_r(\delta) \le \frac{1-r^2}{\eta}$, we have

$$\left|\frac{1}{2\pi}\left(P_r(\delta)(\alpha(\delta+t_0) - L\delta) - P_r(-\delta)(\alpha(-\delta+t_0) + L\delta)\right)\right| \\ \le \frac{1-r^2}{2\pi\eta}\left|\alpha(\delta+t_0) - L\delta - \alpha(-\delta+t_0) - L\delta\right|.$$

Since $(1-r^2)$ approaches 0 as r approaches 1 from below, we can choose $s \in [s_2, 1)$ such that

$$\frac{1-r^2}{2\pi\eta}\left|\alpha(\delta+t_0) - L\delta - \alpha(-\delta+t_0) - L\delta\right| < \frac{\varepsilon}{4}$$

for all $r \in [s, 1)$. Hence

$$\left|\frac{1}{2\pi}\left(P_r(\delta)(\alpha(\delta+t_0) - L\delta) - P_r(-\delta)(\alpha(-\delta+t_0) + L\delta)\right)\right| < \frac{\varepsilon}{4} \text{ for all } r \in [s,1).$$

Define the function $D_r(\psi)$ by

$$D_r(\psi) = \frac{dP_r}{d\psi} = -\frac{(1-r^2)(2r\sin\psi)}{(1-2r\cos\psi + r^2)^2}.$$

Clearly $D_r(\psi) < 0$ for all $r \in [0,1)$ and all $\psi \in [0,\delta]$ (recall that $\delta < \pi$). Since $D_r(\psi) = -D_r(-\psi)$, the change of variables $\psi = -\omega$ yields

$$\begin{aligned}
&\frac{1}{2\pi}\int_{-\delta}^{0}(\alpha(\psi+t_0) - L\psi)\, D_r(\psi)\, d\psi \\
&\quad = \frac{1}{2\pi}\int_{\delta}^{0}(\alpha(-\omega+t_0) + L\omega)\,(-D_r(\omega))(-d\omega) \\
&\quad = -\frac{1}{2\pi}\int_{0}^{\delta}(\alpha(t_0-\omega) + L\omega)\, D_r(\omega) d\omega.
\end{aligned}$$

We now compute

$$\begin{aligned}
&\frac{1}{2\pi}\int_{-\delta}^{\delta}(\alpha(\psi+t_0) - L\psi)\, D_r(\psi)\, d\psi \\
=&\frac{1}{2\pi}\int_{-\delta}^{0}(\alpha(\psi+t_0) - L\psi)\, D_r(\psi)\, d\psi + \frac{1}{2\pi}\int_{0}^{\delta}(\alpha(\psi+t_0) - L\psi)\, D_r(\psi)\, d\psi \\
=&\frac{1}{2\pi}\int_{0}^{\delta}(\alpha(t_0+\psi) - \alpha(t_0-\psi) - 2L\psi)\, D_r(\psi)\, d\psi \\
=&\frac{1}{2\pi}\int_{0}^{\delta}\left(\frac{\alpha(t_0+\psi) - \alpha(t_0-\psi)}{2\psi} - L\right) 2\psi D_r(\psi)\, d\psi.
\end{aligned}$$

Hence

$$\left|\frac{1}{2\pi}\int_{-\delta}^{\delta}(\alpha(\psi+t_0)-L\psi)\,\frac{dP_r}{d\psi}\,d\psi\right|$$
$$\leq \frac{1}{2\pi}\int_0^{\delta}\left|\frac{\alpha(t_0+\psi)-\alpha(t_0-\psi)}{2\psi}-L\right|(-2\psi D_r(\psi))\,d\psi$$
$$\leq \left(\frac{\varepsilon}{8}\right)\frac{1}{2\pi}\int_0^{\delta}(-2\psi D_r(\psi))\,d\psi.$$

An integration by parts gives

$$\int_0^{\delta}\psi D_r(\psi)\,d\psi = \delta P_r(\delta)-\int_0^{\delta}P_r(\psi)\,d\psi,$$

and hence

$$\begin{aligned}\left|\frac{1}{2\pi}\int_{-\delta}^{\delta}(\alpha(\psi+t_0)-L\psi)\,\frac{d}{d\psi}P_r(\psi)\,d\psi\right| &\leq \left(\frac{\varepsilon}{8}\right)\frac{1}{2\pi}\int_0^{\delta}(-2\psi D_r(\psi))\,d\psi\\ &= \left(\frac{\varepsilon}{8}\right)\frac{-2}{2\pi}\left(\delta P_r(\delta)-\int_0^{\delta}P_r(\psi)\,d\psi\right)\\ &\leq \left(\frac{\varepsilon}{8}\right)\frac{1}{2\pi}2\int_0^{\delta}P_r(\psi)\,d\psi\\ &\leq \left(\frac{\varepsilon}{8}\right)\frac{1}{2\pi}2(2\pi)\\ &= \frac{\varepsilon}{4}.\end{aligned}$$

Thus, for $r\in[s,1)$, we have

$$\begin{aligned}\left|u(re^{it_0})-L\right| &\leq \left|\frac{1}{2\pi}\int_{-\delta}^{\delta}P_r(\psi)\,(d\alpha(\psi+t_0)-L\,d\psi)\right|\\ &\quad+\left|\frac{1}{2\pi}\int_{\delta}^{2\pi-\delta}P_r(\psi)\,(d\alpha(\psi+t_0)-L\,d\psi)\right|\\ &\leq \left|\frac{1}{2\pi}\int_{\delta}^{2\pi-\delta}P_r(\psi)\,(d\alpha(\psi+t_0)-L\,d\psi)\right|\\ &\quad+\left|\frac{1}{2\pi}(P_r(\delta)(\alpha(\delta+t_0)-L\delta)-P_r(-\delta)(\alpha(-\delta+t_0)+L\delta))\right|\\ &\quad+\left|\frac{1}{2\pi}\int_{-\delta}^{\delta}(\alpha(\psi+t_0)-L\psi)\,\frac{dP_r}{d\psi}\,d\psi\right|\\ &< \frac{\varepsilon}{2}+\frac{\varepsilon}{4}+\frac{\varepsilon}{4}\\ &= \varepsilon.\end{aligned}$$

Therefore, if $r \in [s, 1)$, we have $|u(re^{it_0}) - L| < \varepsilon$. □

We state the following special case of Fatou's theorem for future reference.

Corollary 1.1.27. *Let ϕ be a function in $\boldsymbol{L}^1(S^1, d\theta)$. Define u by*

$$u(re^{it}) = \frac{1}{2\pi} \int_0^{2\pi} P_r(\theta - t)\phi(e^{i\theta})\, d\theta.$$

Then

$$\lim_{r \to 1^-} u(re^{it})$$

exists for almost all t and equals $\phi(e^{it})$ a.e.

Proof. Define α by

$$\alpha(\theta) = \int_0^{\theta} \phi(e^{ix})\, dx.$$

Then α has bounded variation (it is, in fact, absolutely continuous) and $\alpha'(\theta) = \phi(e^{i\theta})$ a.e. Thus Fatou's theorem (Theorem 1.1.26) gives the result. □

The following corollary is an important application of Fatou's theorem. It is often convenient to identify $\boldsymbol{H}^2$ with $\widetilde{\boldsymbol{H}}^2$; in some contexts, we will refer to f and its boundary function $\tilde{f}$ interchangeably. The next corollary provides further justification for this identification.

Corollary 1.1.28. *If $f \in \boldsymbol{H}^2$, then $\lim_{r \to 1^-} f(re^{i\theta}) = \tilde{f}(e^{i\theta})$ for almost all θ.*

Proof. Recall that if $f \in \boldsymbol{H}^2$, then

$$f(re^{i\theta}) = \frac{1}{2\pi} \int_0^{2\pi} P_r(\theta - t)\tilde{f}(e^{i\theta})\, d\theta$$

(by the Poisson integral formula; see Theorem 1.1.21). Thus the previous corollary yields $\lim_{r \to 1^-} f(re^{i\theta}) = \tilde{f}(e^{i\theta})$ a.e. □

Corollary 1.1.29. *If $f \in \boldsymbol{H}^\infty$, then $\tilde{f} \in \boldsymbol{L}^\infty$.*

Proof. It follows from the above corollary (Corollary 1.1.28) that the essential supremum of $\tilde{f}$ is at most $\|f\|_\infty$. □

Definition 1.1.30. The space $\widetilde{\boldsymbol{H}}^\infty$ is defined to be $\widetilde{\boldsymbol{H}}^2 \bigcap \boldsymbol{L}^\infty$.

The notation $\widetilde{\boldsymbol{H}}^\infty$ is justified since, by Corollaries 1.1.24 and 1.1.29, f is in $\boldsymbol{H}^\infty$ if and only $\tilde{f}$ is in $\widetilde{\boldsymbol{H}}^\infty$.

1.2 Some Facts from Functional Analysis

In this section we introduce some basic facts from functional analysis that we will use throughout this book. We require the fundamental properties of bounded linear operators on Hilbert spaces. When we use the term "bounded linear operator", we mean a bounded linear operator taking a Hilbert space into itself (although many of the definitions and theorems below apply to bounded linear operators on arbitrary Banach spaces). When talking about operators on an arbitrary Hilbert space, we use $\mathcal{H}$ to denote the Hilbert space.

The spectrum of an operator is one of the fundamental concepts in operator theory.

Definition 1.2.1. If A is a bounded linear operator on a Hilbert space $\mathcal{H}$, the *spectrum* of A, denoted by $\sigma(A)$, is the set of all complex numbers λ such that $A - \lambda$ is not invertible. (The notation $A - \lambda$ is shorthand for $A - \lambda I$, where I is the identity operator on $\mathcal{H}$.)

Definition 1.2.2. Let A be a bounded linear operator. The *spectral radius* of A, denoted by $r(A)$, is

$$r(A) = \sup\{|\lambda| : \lambda \in \sigma(A)\}.$$

As we note below, the spectrum is nonempty and bounded, and thus the spectral radius is well-defined. Various parts of the spectrum are important.

Definition 1.2.3. The complex number λ is an *eigenvalue* of the bounded operator A if $Af = \lambda f$ for some nonzero f; the vector f is then said to be an *eigenvector* of A. The set of all eigenvalues of A is called the *point spectrum* of A and is denoted by $\Pi_0(A)$. The *approximate point spectrum* is the set $\Pi(A)$ of complex numbers λ such that there exists a sequence $\{f_n\}$ of unit vectors satisfying $\{\|(A - \lambda)f_n\|\} \to 0$ as $n \to \infty$.

The following properties of spectra are very elementary and very well known.

Theorem 1.2.4. *Let A be a bounded linear operator.*

(i) If $\|1 - A\| < 1$, then A is invertible.

(ii) The spectrum of A is a nonempty compact subset of $\mathbb{C}$.

(iii) If A is an invertible operator, then

$$\sigma(A^{-1}) = \left\{ \frac{1}{\lambda} : \lambda \in \sigma(A) \right\}.$$

(iv) If A^* denotes the Hilbert space adjoint of A, then

$$\sigma(A^*) = \{ \overline{\lambda} : \lambda \in \sigma(A) \}.$$

(v) The spectral radius formula holds:

$$r(A) = \lim_{n \to \infty} \|A^n\|^{1/n}.$$

In particular, $r(A) \leq \|A\|$.

(vi) If A is an operator on a finite-dimensional space, then $\sigma(A) = \Pi_0(A)$ (for operators on infinite-dimensional spaces, $\Pi_0(A)$ may be the empty set).

(vii) The number λ is in $\Pi(A)$ if and only if $A - \lambda$ is not bounded below; i.e., there is no constant $c > 0$ such that $\|(A - \lambda)f\| \geq c\|f\|$ for all $f \in \mathcal{H}$. Moreover, $A - \lambda$ is bounded below if and only if $A - \lambda$ is injective and the range of $A - \lambda$ is closed. In particular, $\Pi_0(A) \subset \Pi(A)$ and $\Pi(A) \subset \sigma(A)$.

Proof. Proofs of the above assertions can be found in most introductory functional analysis textbooks. In particular, see [12, pp. 195–198], [42, pp. 188–194], [48, pp. 252–255], and [55, Chapter V]. □

The following part of the spectrum is not quite as widely studied as those mentioned above.

Definition 1.2.5. The *compression spectrum*, denoted by $\Gamma(A)$, is the set of complex numbers λ such that $A - \lambda$ does not have dense range.

Theorem 1.2.6. *For every bounded linear operator A, $\sigma(A) = \Pi(A) \cup \Gamma(A)$.*

Proof. Clearly both $\Pi(A)$ and $\Gamma(A)$ are contained in $\sigma(A)$. If λ is not in $\Pi(A)$, it follows that $A - \lambda$ is bounded below, and hence that $A - \lambda$ is one-to-one and has closed range. If, in addition, λ is not in $\Gamma(A)$, then $A - \lambda$ has dense range. But if $A - \lambda$ has closed range and dense range, then $A - \lambda$ maps onto $\mathcal{H}$. Since $A - \lambda$ is also injective, this implies that $A - \lambda$ is invertible; i.e., λ is not in $\sigma(A)$. □

Theorem 1.2.7. *For every bounded linear operator A, the boundary of $\sigma(A)$ is contained in $\Pi(A)$. In particular, $\Pi(A)$ is nonempty.*

Proof. Let λ be in the boundary of $\sigma(A)$ and assume that $\lambda \notin \Pi(A)$. Choose a sequence $\{\lambda_n\} \to \lambda$ such that $\lambda_n \notin \sigma(A)$ for every natural number n.

We claim that there exists a constant $k > 0$ and a positive integer M such that

$$\|(A - \lambda_n)f\| \geq k\|f\| \qquad \text{for all } f \in \mathcal{H} \text{ whenever } n \geq M.$$

If this were false, then for every $\varepsilon > 0$ and every natural M there would exist an $n \geq M$ and an f_n of norm 1 such that

$$\|(A - \lambda_n)f_n\| < \frac{\varepsilon}{2}.$$

Given any $\varepsilon > 0$, choose a natural number M such that $|\lambda_n - \lambda| < \frac{\varepsilon}{2}$ for all $n \geq M$. With this ε and M, choose n and f_n as above. Then

$$\|(A - \lambda)f_n\| \leq \|(A - \lambda_n)f_n\| + \|(\lambda_n - \lambda)f_n\| < \varepsilon.$$

But this would imply that $\lambda \in \Pi(A)$, which would be a contradiction, so the claim is proved.

To contradict the assumption $\lambda \notin \Pi(A)$, we now show that $\lambda \notin \Gamma(A)$ (since $\lambda \in \sigma(A) = \Pi(A) \cup \Gamma(A)$, this is a contradiction). Choose any vector g different from 0. We must show that g is in the closure of the range of $(A-\lambda)$. Given $\varepsilon > 0$, we can choose N sufficiently large (in fact, larger than M) such that if $n \geq N$ then

$$|\lambda_n - \lambda| < \frac{k}{\|g\|}\,\varepsilon.$$

Since $\lambda_n \notin \sigma(A)$, there exists $f_n \in \mathcal{H}$ with $(A-\lambda_n)f_n = g$. The claim then implies that $\|g\| \geq k\|f_n\|$ for all $n \geq N$. Then

$$\begin{aligned}
\|(A - \lambda)f_n - g\| &= \|\,((A - \lambda_n)f_n - g) + (\lambda_n - \lambda)f_n\| \\
&= |\lambda_n - \lambda|\,\|f_n\| \\
&\leq \frac{1}{k}\|g\|\,|\lambda_n - \lambda| \\
&< \varepsilon.
\end{aligned}$$

Hence g is in the closure of the range of $(A - \lambda)$. Thus $\lambda \notin \Gamma(A)$. □

The numerical range of an operator is not as important as the spectrum, but it is very useful in several contexts.

Definition 1.2.8. The *numerical range* of A, denoted by $W(A)$, is the following subset of the complex plane:

$$\{(Af, f) \,:\, f \in \mathcal{H}, \quad \|f\| = 1\}.$$

The most fundamental property of the numerical range is the following.

Theorem 1.2.9 (Toeplitz–Hausdorff Theorem). *The numerical range of a bounded linear operator is a convex subset of the complex plane.*

Proof. There are several well-known elementary proofs of this theorem (cf. [27, p. 113], [80]). □

Example 1.2.10. *If A is a finite diagonal matrix*

$$A = \begin{pmatrix} d_1 & 0 & 0 & \cdots & 0 \\ 0 & d_2 & 0 & \cdots & 0 \\ 0 & 0 & d_3 & \cdots & 0 \\ \vdots & \vdots & \vdots & \ddots & \vdots \\ 0 & 0 & \cdots & \cdots & d_n \end{pmatrix},$$

then $W(A)$ is the convex hull of $\{d_1, d_2, \ldots, d_n\}$.

Proof. If $f = (f_1, f_2, f_3, \ldots, f_n)$, then $(Af, f) = \sum_{i=1}^{n} d_i |f_i|^2$. □

Thus, in the case of finite diagonal matrices, the numerical range is the convex hull of the spectrum of A.

Theorem 1.2.11. *For every operator A, $\sigma(A) \subset \overline{W(A)}$ (i.e., the closure of the numerical range).*

Proof. As was mentioned above, $\sigma(A) = \Pi(A) \cup \Gamma(A)$ (by Theorem 1.2.6). We first prove that $\Pi(A) \subset \overline{W(A)}$. Let $\lambda \in \Pi(A)$. Then there exists a sequence $\{f_n\}$ in $\mathcal{H}$ such that $\|f_n\| = 1$ for all n and $\{\|(A - \lambda)f_n\|\} \to 0$ as $n \to \infty$. But then

$$|(Af_n, f_n) - \lambda| = |\ (Af_n, f_n) - \lambda(f_n, f_n)\ | = |\ ((A - \lambda)f_n, f_n)\ | \leq \|(A-\lambda)f_n\|.$$

This implies that, as $n \to \infty$, $\{(Af_n, f_n)\} \to \lambda$; i.e., $\lambda \in \overline{W(A)}$. Therefore $\Pi(A) \subset \overline{W(A)}$.

Now we prove that $\Gamma(A) \subset W(A)$. Let $\lambda \in \Gamma(A)$. Since $A - \lambda$ does not have dense range, it follows that there exists a nonzero vector $g \in \mathcal{H}$ with $\|g\| = 1$ such that g is orthogonal to $(A - \lambda)f$ for all $f \in \mathcal{H}$. That is, for all $f \in \mathcal{H}$, $((A - \lambda)f, g) = 0$. In particular, taking f to be g yields

$$0 = ((A - \lambda)g, g) = (Ag, g) - \lambda(g, g) = (Ag, g) - \lambda.$$

Thus $(Ag, g) = \lambda$; i.e., $\lambda \in W(A)$. This concludes the proof. □

The following is a generalization of Example 1.2.10.

Theorem 1.2.12. *If A is normal, then $\overline{W(A)}$ (the closure of the numerical range of A) is the convex hull of $\sigma(A)$.*

Proof. By one form of the spectral theorem ([12, p. 272], [41, p. 13], [42, p. 246]), we may assume that A is multiplication by an $\boldsymbol{L}^\infty(X, d\mu)$ function ϕ acting on a space $\boldsymbol{L}^2(X, d\mu)$ for some measurable subset X of the complex plane and some measure $d\mu$ on it.

We know that $\sigma(A) \subset \overline{W(A)}$ by the previous theorem. By the Toeplitz–Hausdorff theorem (Theorem 1.2.9), it follows that the convex hull of $\sigma(A)$ is also contained in $\overline{W(A)}$.

To prove the opposite inclusion, it suffices to prove that every closed half-plane in $\mathbb{C}$ that contains $\sigma(A)$ also contains $W(A)$. By rotation and translation, assume that $\sigma(A)$ is contained in the right-hand plane $\operatorname{Re} z \geq 0$. We need only show that $W(A)$ is contained in this half-plane.

But if $A = M_\phi$ is multiplication by ϕ, then $\sigma(A) = \sigma(M_\phi)$ is the essential range of the function ϕ (this fact is discussed in the special context of Lebesgue-measurable functions on S^1 in Theorem 3.1.6 below; the proof in the general situation is exactly the same as in that special case). It follows that $\operatorname{Re} \phi \geq 0$ almost everywhere. Therefore

$$(Af, f) = (M_\phi f, f) = \int_X \phi \; |f|^2 d\mu.$$

Thus $\operatorname{Re}(Af, f) \geq 0$. $\square$

We will be considering invariant subspaces of various operators.

Definition 1.2.13. By a *subspace* of a Hilbert space, we mean a subset of the space that is closed in the topological sense in addition to being closed under the vector space operations. By a *linear manifold* we mean a subset that is closed under the vector operations but is not necessarily closed in the topology.

We will often have occasion to consider the smallest subspace containing a given collection of vectors.

Definition 1.2.14. If $\mathcal{S}$ is any nonempty subset of a Hilbert space, then the *span of* $\mathcal{S}$, often denoted by

$$\bigvee\{f : f \in \mathcal{S}\} \quad \text{or} \quad \bigvee \mathcal{S},$$

is the intersection of all subspaces containing $\mathcal{S}$. It is obvious that $\bigvee \mathcal{S}$ is always a subspace.

Definition 1.2.15. If A is an operator and $\mathcal{M}$ is a subspace, we say that $\mathcal{M}$ is an *invariant subspace* of A if $A\mathcal{M} \subset \mathcal{M}$. That is, $\mathcal{M}$ is invariant under A if $f \in \mathcal{M}$ implies $Af \in \mathcal{M}$.

The trivial subspaces, $\{0\}$ and $\mathcal{H}$, are invariant under every operator. One of the most famous unsolved problems in analysis (the *invariant subspace problem*) is the question whether every bounded linear operator on an infinite-dimensional Hilbert space has a nontrivial invariant subspace.

Notation 1.2.16. If $\mathcal{M}$ is an invariant subspace of the operator A, then $A\big|_{\mathcal{M}}$ is the restriction of the operator A to $\mathcal{M}$.

Definition 1.2.17. Given a vector f and a bounded linear operator A, *the invariant subspace generated by* f is the subspace

$$\bigvee_{n=0}^{\infty} \{A^n f\}.$$

We say that an invariant subspace $\mathcal{M}$ of A is *cyclic* if there is a vector g such that $\mathcal{M} = \bigvee_{n=0}^{\infty}\{A^n g\}$. If

$$\bigvee_{n=0}^{\infty} \{A^n g\} = \mathcal{H},$$

we say that g is a *cyclic vector for* A.

Clearly, the invariant subspace problem can be rephrased: does every bounded linear operator on Hilbert space have a noncyclic vector other than zero?

It turns out that the collection of subspaces invariant under an operator (or any family of operators) is a lattice.

Definition 1.2.18. A *lattice* is a partially ordered set in which every pair of elements has a least upper bound and a greatest lower bound. A lattice is *complete* if every nonempty subset of the lattice has a least upper bound and a greatest lower bound.

It is easily seen that the collection of all subspaces invariant under a given bounded linear operator is a complete lattice under inclusion, where the least upper bound of a subcollection is its span and the greatest lower bound of a subcollection is its intersection.

Notation 1.2.19. For A a bounded linear operator, we use the notation $\operatorname{Lat} A$ to denote the lattice of all invariant subspaces of A.

Theorem 1.2.20. *Let A be a bounded linear operator. Then $\mathcal{M} \in \operatorname{Lat} A$ if and only if $\mathcal{M}^\perp \in \operatorname{Lat} A^*$.*

Proof. This follows immediately from the fact that, for $f \in \mathcal{M}$ and $g \in \mathcal{M}^\perp$, $(Af, g) = (f, A^*g)$. □

Recall that, if $\mathcal{M}$ is a subspace of $\mathcal{H}$, every vector $f \in \mathcal{H}$ can be written uniquely in the form $f = m + n$, where $m \in \mathcal{M}$ and $n \in \mathcal{M}^\perp$.

Notation 1.2.21. If $\mathcal{M}$ and $\mathcal{N}$ are subspaces of a Hilbert space, the notation $\mathcal{M} \oplus \mathcal{N}$ is used to denote $\{m + n : m \in \mathcal{M} \text{ and } n \in \mathcal{N}\}$ when every vector in $\mathcal{M}$ is orthogonal to every vector in $\mathcal{N}$. The expression $\mathcal{M} \ominus \mathcal{N}$ denotes $\mathcal{M} \cap \mathcal{N}^\perp$.

Definition 1.2.22. If $\mathcal{M}$ is a subspace then the *projection* onto $\mathcal{M}$ is the operator defined by $Pf = g$, where $f = g + h$ with $g \in \mathcal{M}$ and $h \in \mathcal{M}^\perp$.

It is easy to see that every projection is a bounded self-adjoint operator of norm at most one. Also, since $P\mathcal{H} = \mathcal{M}$, $P\mathcal{H}$ is always a subspace.

Theorem 1.2.23. *If $\mathcal{M} \in \operatorname{Lat} A$ and P is the projection onto $\mathcal{M}$, then $AP = PAP$. Conversely, if P is a projection and $AP = PAP$, then $P\mathcal{H} \in \operatorname{Lat} A$.*

Proof. Let $\mathcal{M} \in \operatorname{Lat} A$ and P be the projection onto $\mathcal{M}$. If $f \in \mathcal{H}$ then $Pf \in \mathcal{M}$ and therefore APf is contained in $A\mathcal{M}$. Since $A\mathcal{M} \subset \mathcal{M}$ it follows that $P(APf) = APf$.

Conversely, let P be a projection and assume that $AP = PAP$. If $f \in P\mathcal{H}$, then $Pf = f$ and therefore $APf = PAPf$ simplifies to $Af = PAf$. Thus $Af \in P\mathcal{H}$ and $P\mathcal{H} \in \operatorname{Lat} A$. □

Recall that a decomposition of a Hilbert space $\mathcal{H}$ in the form $\mathcal{M} \oplus \mathcal{M}^\perp$ leads to a block matrix representation of operators on $\mathcal{H}$. If P is the projection of $\mathcal{H}$ onto $\mathcal{M}$ and A_1 is the restriction of PA to $\mathcal{M}$, A_2 is the restriction of PA to $\mathcal{M}^\perp$, A_3 is the restriction of $(I - P)A$ to $\mathcal{M}$, and A_4 is the restriction of $(I - P)A$ to $\mathcal{M}^\perp$, then A can be represented as

$$A = \begin{pmatrix} A_1 & A_2 \\ A_3 & A_4 \end{pmatrix}.$$

with respect to the decomposition $\mathcal{M}\oplus\mathcal{M}^\perp$. That is, if $f = g+h$ with $g \in \mathcal{M}$ and $h \in \mathcal{M}^\perp$, we have

$$Af = \begin{pmatrix} A_1 & A_2 \\ A_3 & A_4 \end{pmatrix} \begin{pmatrix} g \\ h \end{pmatrix} = \begin{pmatrix} A_1 g + A_2 h \\ A_3 g + A_4 h \end{pmatrix} = (A_1 g + A_2 h) + (A_3 g + A_4 h).$$

If the subspace $\mathcal{M}$ is invariant under A, then Theorem 1.2.23 implies that $A_3 = 0$. Thus each nontrivial invariant subspace of A yields an upper triangular representation of A.

Definition 1.2.24. The subspace $\mathcal{M}$ *reduces* the operator A if both $\mathcal{M}$ and $\mathcal{M}^\perp$ are invariant under A.

Theorem 1.2.25. *Let P be the projection onto the subspace $\mathcal{M}$. Then $\mathcal{M}$ is a reducing subspace for A if and only if $PA = AP$. Also, $\mathcal{M}$ reduces A if and only if $\mathcal{M}$ is invariant under both A and A^*.*

Proof. If $\mathcal{M}$ is a reducing subspace, then $\mathcal{M}$ and $\mathcal{M}^\perp$ are invariant under A. If P is the projection onto $\mathcal{M}$, it is easily seen that $I - P$ is the projection onto $\mathcal{M}^\perp$. The previous theorem then implies $A(I-P) = (I-P)A(I-P)$. Expanding the latter equation gives $A - AP = A - PA - AP + PAP$, which simplifies to $PAP = PA$. Since $\mathcal{M} \in \operatorname{Lat} A$ we also have that $AP = PAP$ and thus $PA = AP$.

Conversely, assume $AP = PA$. Let $f \in \mathcal{M}$; to prove $\mathcal{M}$ is invariant we need to show that $Af \in \mathcal{M}$. By hypothesis, $PAf = APf$ and, since $Pf = f$, it follows that $PAf = Af$, which is equivalent to $Af \in \mathcal{M}$. Thus $\mathcal{M} \in \operatorname{Lat} A$. We also have that $(I-P)A = A(I-P)$ and thus an analogous argument shows that if $f \in \mathcal{M}^\perp$, then $Af \in \mathcal{M}^\perp$. Hence $\mathcal{M}^\perp \in \operatorname{Lat} A$ and A is reducing.

For the second part of the theorem notice that, since P is self-adjoint, $PA = AP$ if and only if $PA^* = A^*P$. This means that $\mathcal{M}$ is reducing for A if and only if $\mathcal{M}$ is reducing for A^*. In particular, $\mathcal{M}$ is invariant for both A and A^*.

For the converse of the second part, observe that $PAP = AP$ and $PA^*P = A^*P$. If we take the adjoint of the latter equation it follows that $AP = PA$ and thus $\mathcal{M}$ is reducing, by the first part of the theorem. □

It is easily seen that the subspace $\mathcal{M}$ reduces A if and only if the decomposition of A with respect to $\mathcal{M} \oplus \mathcal{M}^\perp$ has the form

$$A = \begin{pmatrix} A_1 & 0 \\ 0 & A_4 \end{pmatrix},$$

where A_1 is an operator on $\mathcal{M}$ and A_4 is an operator on $\mathcal{M}^\perp$. This matrix representation shows why the word "reducing" is used.

Definition 1.2.26. The *rank* of the operator A is the dimension of its range.

Finite-rank operators (i.e., those operators whose rank is a natural number) share many properties with operators on finite-dimensional spaces and thus are particularly tractable. Operators whose rank is 1 are often very useful.

Notation 1.2.27. Given vectors f and $g \in \mathcal{H}$, we define the operator $f \otimes g$ mapping $\mathcal{H}$ into itself by $(f \otimes g)h = (h, g)f$.

Note that if neither f nor g is zero, the operator $f \otimes g$ has rank 1 since its range consists of multiples of f. Clearly, $f \otimes g = 0$ if and only if either $f = 0$ or $g = 0$.

Theorem 1.2.28. *(i) If A is an operator of rank* 1*, then there exist f and g in $\mathcal{H}$ with $A = f \otimes g$.*

(ii) $\|f \otimes g\| = \|f\| \, \|g\|$.

(iii) For bounded operators A and B, $A(f \otimes g)B = (Af) \otimes (B^*g)$.

(iv) Two nonzero rank-one operators $f_1 \otimes g_1$ and $f_2 \otimes g_2$ are equal if and only if there exists a complex number c other than 0 *such that $f_1 = cf_2$ and $g_2 = \overline{c} g_1$.*

Proof of (i): Let f be any nonzero vector in the range of A. Since the range of A is one-dimensional, there is a bounded linear functional λ such that $Ah = \lambda(h)f$ for all vectors h. By the Riesz representation theorem ([12, p. 13], [28, pp. 31–32], [55, p. 142]), there is a g in $\mathcal{H}$ such that $\lambda(h) = (h, g)$ for all h in $\mathcal{H}$. Therefore $Ah = (h, g)f = (f \otimes g)h$ for all h. □

Proof of (ii): Let $h \in \mathcal{H}$. Then

$$\|(f \otimes g)h\| = \|(h, g)f\| \leq \|h\| \, \|g\| \, \|f\|.$$

Taking the supremum over all h with $\|h\| = 1$ gives $\|f \otimes g\| \leq \|f\| \, \|g\|$. To establish the reverse inequality, observe that, for $g \neq 0$,

$$\left\| (f \otimes g) \frac{g}{\|g\|} \right\| = \left\| \left(\frac{g}{\|g\|}, g \right) f \right\| = \|g\| \, \|f\|.$$

Thus $\|f \otimes g\| \geq \|f\| \, \|g\|$. □

Proof of (iii): Let A and B be bounded operators. If $h \in \mathcal{H}$, then

$$\begin{aligned}(A(f \otimes g)B)(h) &= (A(f \otimes g))(Bh) = A((Bh, g)f) \\ &= (Bh, g)Af = (h, B^*g)Af \\ &= ((Af) \otimes (B^*g))(h).\end{aligned}$$

□

Proof of (iv): Assume the equality of the rank-one operators. Since $f_1 \otimes g_1$ and $f_2 \otimes g_2$ are both nonzero, none of the four vectors involved are 0. Note that

$$(f_1 \otimes g_1)\frac{g_1}{\|g_1\|^2} = f_1 \quad \text{and} \quad (f_2 \otimes g_2)\frac{g_1}{\|g_1\|^2} = \frac{(g_1, g_2)}{\|g_1\|^2} f_2,$$

and thus $f_1 = cf_2$, where $c = \frac{(g_1, g_2)}{\|g_1\|^2}$. Since $f_1 = cf_2$, we have $(cf_2) \otimes g_1 = f_2 \otimes g_2$. Thus, for all $h \in \mathcal{H}$,

$$(h, g_1)cf_2 = (h, g_2)f_2.$$

This implies that $(h, \overline{c}g_1) = (h, g_2)$ for all h, so $\overline{c}g_1 = g_2$.

For the converse, note that, for every $h \in \mathcal{H}$,

$$\begin{aligned}(f_1 \otimes g_1)(h) &= ((cf_2) \otimes g_1)(h) = (h, g_1)cf_2 \\ &= (h, \overline{c}g_1)f_2 = (f_2 \otimes (\overline{c}g_1))(h) \\ &= (f_2 \otimes g_2)(h).\end{aligned}$$

Therefore the rank-one operators are equal. □

1.3 Exercises

1.1. Show that $\boldsymbol{H}^\infty$ is a Banach space under the $\|\cdot\|_\infty$ norm.

1.2. Find a function in $\boldsymbol{H}^2$ whose square is not in $\boldsymbol{H}^2$.

1.3. Prove that the only real-valued functions in $\widetilde{\boldsymbol{H}}^2$ are the constants.

1.4. Show that the only functions in $\widetilde{\boldsymbol{H}}^2$ whose conjugates are also in $\widetilde{\boldsymbol{H}}^2$ are the constants.

1.5. Prove that $\left(\frac{1}{1-z}\right)^{1/2}$ is not in $\boldsymbol{H}^2$.

1.6. Show that a function f analytic on $\mathbb{D}$ is in $\boldsymbol{H}^2$ if there is a harmonic function u on $\mathbb{D}$ such that $|f(z)|^2 \leq u(z)$ for all $z \in \mathbb{D}$. (Such a function u is said to be a *harmonic majorant of the function* $|f(z)|^2$.) The converse of this fact is stated below in Exercise 2.12.

1.7. Define $\widetilde{\boldsymbol{H}}^1$ to be the set of all functions in $\boldsymbol{L}^1(S^1)$ whose Fourier coefficients corresponding to negative indices are zero. Prove that the product of two functions in $\widetilde{\boldsymbol{H}}^2$ is in $\widetilde{\boldsymbol{H}}^1$.

1.8. Let u be a real-valued function in $\boldsymbol{L}^2$. Show that there exists a real-valued function v in $\boldsymbol{L}^2$ such that $u + iv$ is in $\widetilde{\boldsymbol{H}}^2$.

1.9. Let f be an even function of a real variable defined in a neighborhood of 0. Show that f has symmetric derivative 0 at 0. Note that this implies that there exist functions that are not even left or right continuous at a point but nonetheless have symmetric derivatives at that point.

1.10. Let A be a bounded linear operator and p be a polynomial. Prove that $\sigma(p(A)) = \{p(z) : z \in \sigma(A)\}$.

1.11. Suppose that a bounded linear operator A has an upper triangular matrix with respect to an orthonormal basis $\{e_n\}_{n=0}^\infty$. Show that every element (Ae_n, e_n) of the diagonal is an eigenvalue of A.

1.12. Let A be a bounded linear operator and λ be a complex number with $|\lambda| = \|A\|$. Prove that λ is in the numerical range of A if and only if λ is an eigenvalue of A.

1.13. Show that the restriction of a normal operator to an invariant subspace is normal if and only if the subspace is reducing.

1.14. Assume that there is a bounded operator A with the following property: there exist subspaces $\mathcal{N}$ and $\mathcal{M}$ invariant under A such that $\mathcal{N} \subset \mathcal{M}$, the dimension of $\mathcal{M} \ominus \mathcal{N}$ is greater than 1, and the only subspaces $\mathcal{L}$ invariant under A that satisfy $\mathcal{N} \subset \mathcal{L} \subset \mathcal{M}$ are $\mathcal{L} = \mathcal{N}$ and $\mathcal{L} = \mathcal{M}$. Show that this assumption implies the existence of an operator B on a Hilbert space whose only invariant subspaces are $\{0\}$ and the entire space.

1.15. Show that the operator A has a nontrivial invariant subspace if and only if the operator equation $XAX = AX$ has a solution other than zero and the identity.

1.16. Show that every operator of finite rank can be written in the form

$$\sum_{k=1}^{n} f_k \otimes g_k$$

for vectors $\{f_1, f_2, \ldots, f_n, g_1, g_2, \ldots, g_n\}$.

1.17. For a bounded operator A on $\boldsymbol{H}^2$, define its *Berezin symbol* as the function $\widetilde{A}$ on $\mathbb{D}$ given by

$$\widetilde{A}(z) = (A\widehat{k_z}, \widehat{k_z}),$$

where $\widehat{k_z} = \frac{k_z}{\|k_z\|}$ is the normalized reproducing kernel. Show that

$$\lim_{|z| \to 1^-} \widetilde{A}(z) = 0$$

for every compact operator A.

1.18. Suppose that $\{A_n\}$ is a sequence of bounded operators such that $\{A_n f\}$ converges for every vector f. Prove that $\{\|A_n\|\}$ is bounded.

1.19. Prove that $\{\|A_n C\|\}$ converges to 0 whenever C is a compact operator and $\{A_n\}$ is a sequence of bounded operators such that $\{A_n f\}$ converges to 0 for all vectors f. (Hint: This can be established by first proving it in the case that C has finite rank, then uniformly approximating any given compact operator by a sequence of finite-rank operators and using the previous exercise to obtain the result as stated.)

1.20. The *Bergman space* is the collection of all functions f analytic on the disk such that $|f(z)|^2$ is integrable with respect to normalized Lebesgue area measure on the disk (i.e., the measure of $\mathbb{D}$ is 1). The inner product of the functions f and g is defined by

$$(f,g) = \frac{1}{\pi}\int_{\mathbb{D}} f(x+iy)\overline{g(x+iy)}\,dx\,dy.$$

(i) Show that the Bergman space is a Hilbert space (i.e., prove the completeness of the inner product space defined above).

(ii) Show that the collection of functions $\{\sqrt{n+1}\,z^n\}_{n=0}^{\infty}$ forms an orthonormal basis for the Bergman space.

(iii) Let M_z be the operator defined by $(M_z f)(z) = zf(z)$. Show that M_z is a bounded linear operator mapping the Bergman space into itself and find the matrix of M_z with respect to the orthonormal basis $\{\sqrt{n+1}z^n\}_{n=0}^{\infty}$.

1.21. The *Dirichlet space* is the collection of all functions f analytic on the disk such that $|f'(z)|^2$ is integrable with respect to normalized Lebesgue area measure on the disk. The inner product of the functions f and g is defined by

$$(f,g) = f(0)\overline{g(0)} + \frac{1}{\pi}\int_{\mathbb{D}} f'(x+iy)\overline{g'(x+iy)}\,dx\,dy.$$

(i) Show that the Dirichlet space is a Hilbert space (i.e., prove the completeness of the inner product space defined above).

(ii) Find an orthonormal basis for the Dirichlet space consisting of monomials in z.

(iii) Let M_z be the operator defined by $(M_z f)(z) = zf(z)$. Show that M_z is a bounded linear operator mapping the Dirichlet space into itself and find the matrix of M_z with respect to the orthonormal basis found in the answer to *(ii)*.

1.4 Notes and Remarks

For each real number $p \geq 1$, the space $\boldsymbol{H}^p$ is defined to consist of the set of all functions f analytic on $\mathbb{D}$ such that

$$\sup_{0<r<1} \frac{1}{2\pi} \int_0^{2\pi} \left| f(re^{i\theta}) \right|^p d\theta < \infty.$$

The $\boldsymbol{H}^p$ norm of f is defined to be

$$\left(\sup_{0<r<1} \frac{1}{2\pi} \int_0^{2\pi} \left| f(re^{i\theta}) \right|^p d\theta \right)^{1/p}.$$

Note that Theorem 1.1.12 above shows that, in the case $p = 2$, this definition is equivalent to the one we have used.

The "$\boldsymbol{H}$" in $\boldsymbol{H}^p$ is in honor of G.H. Hardy, a contributor to the fundamentals of the subject. Duren [17] suggests that the historical starting point of the theory of $\boldsymbol{H}^p$ spaces is Hardy's paper [98].

All the $\boldsymbol{H}^p$ spaces are Banach spaces; $\boldsymbol{H}^2$ is the only one that is a Hilbert space. Many of the results in this chapter hold for all $\boldsymbol{H}^p$ spaces; however, the proofs are often easier in the case of $\boldsymbol{H}^2$ than in the general case. Good introductions to $\boldsymbol{H}^p$ spaces include the books by Duren [17], Hoffman [32], Koosis [33], and Chapter 17 of Rudin [47].

Exercises 1.20 and 1.21 give the definitions of the Bergman and Dirichlet spaces. These spaces, and also Hardy, Bergman, and Dirichlet spaces of functions of several variables, have been extensively studied; see, for example, Cowen and MacCluer [14], Duren and Schuster [18], Hedenmalm, Korenblum and Zhu [31], Rudin [45], and Zhu [57].

Example 1.1.14 is due to Nordgren [121]. The proof presented in the text is Jaime Cruz-Sampedro's simplification of our previous simplification of Nordgren's proof.

Fatou's theorem (Theorem 1.1.26) on the boundary values of analytic functions is one of the earliest theorems in this subject [92].

Many of the basic properties of Hilbert space were discovered by Hilbert [105]; the theory has been extensively developed by many mathematicians. Good introductions to the theory of operators on Hilbert space include Akhiezer and Glazman [2], Conway [12,13], Halmos [27], Reed and Simon [42], and Rudin [48].

The question of the existence of nontrivial invariant subspaces for bounded linear operators, the invariant subspace problem, goes back at least to John von Neumann. It has been shown, by Enflo [89] and Read [142, 143], that there are operators on Banach spaces without nontrivial invariant subspaces. However, it is still unknown whether every bounded linear operator on an infinite-dimensional Hilbert space has a nontrivial invariant subspace. There are a number of affirmative results under various hypothesis; see [41].

Exercise 1.6 gives an alternative approach to the definition of $\boldsymbol{H}^2$ that can be used to define analogous spaces consisting of functions analytic on other domains; see Duren [17, Chapter 10]. Exercise 1.15 is a lemma in the approach by Aronszajn and Smith [60] to establishing the existence of nontrivial invariant subspaces for compact operators. Exercise 1.18 is a special case of the principle of uniform boundedness; see Conway [12, p. 95] or Rudin [48, p. 43]. A solution to Exercise 1.12 can be found in [27, Solution 212].

Chapter 2

The Unilateral Shift and Factorization of Functions

We introduce the unilateral shift, one of the most interesting operators. The study of the invariant subspaces of this operator leads naturally to a factorization of functions in $\boldsymbol{H}^2$.

2.1 The Shift Operators

Definition 2.1.1. On ℓ^2, we define the *unilateral shift operator* U by

$$U(a_0, a_1, a_2, a_3, \dots) = (0, a_0, a_1, a_2, a_3, \dots)$$

for $(a_0, a_1, a_2, a_3, \dots) \in \ell^2$.

Theorem 2.1.2. *(i) The unilateral shift is an isometry (i.e., $\|Uf\| = \|f\|$ for all $f \in \ell^2$).*

(ii) The adjoint, U^*, of the unilateral shift has the following form:

$$U^*(a_0, a_1, a_2, a_3, \dots) = (a_1, a_2, a_3, \dots)$$

for $(a_0, a_1, a_2, a_3, \dots) \in \ell^2$. (The operator U^* is the backward unilateral shift.*)*

Proof. To prove *(i)*, we must show that $\|(a_0, a_1, a_2, \dots)\| = \|(0, a_0, a_1, a_2, \dots)\|$. But this is trivial since $\sum_{k=0}^{\infty} |a_k|^2 = |0|^2 + \sum_{k=1}^{\infty} |a_{k-1}|^2$.

To prove *(ii)*, let A be the operator defined by $A(a_0, a_1, a_2, a_3, \dots) = (a_1, a_2, a_3, a_4, \dots)$. Let $x = (a_0, a_1, a_2, \dots)$ and $y = (b_0, b_1, b_2, \dots)$ be any two vectors. Notice that

$$(Ux, y) = ((0, a_0, a_1, a_2, \dots), (b_0, b_1, b_2, b_3, \dots)) = \sum_{k=1}^{\infty} a_{k-1}\overline{b_k}$$

and

$$(x, Ay) = ((a_0, a_1, a_2, a_3 \dots), (b_1, b_2, b_3, b_4, \dots)) = \sum_{k=0}^{\infty} a_k\overline{b_{k+1}}.$$

Since these sums are equal, it follows that $A = U^*$. □

There are also bilateral shifts, defined on two-sided sequences.

Definition 2.1.3. The space $\ell^2(\mathbb{Z})$ is defined as the space of all two-sided square-summable sequences; that is,

$$\ell^2(\mathbb{Z}) = \left\{ (\dots, a_{-2}, a_{-1}, \boldsymbol{a_0}, a_1, a_2, \dots) : \sum_{n=-\infty}^{\infty} |a_n|^2 < \infty \right\}.$$

Note that the zeroth coordinate of the sequence is written in boldface; this is necessary in order to distinguish a sequence from a shift of itself.

Definition 2.1.4. The *bilateral shift* is the operator W on $\ell^2(\mathbb{Z})$ defined by

$$W(\dots, a_{-2}, a_{-1}, \boldsymbol{a_0}, a_1, a_2, \dots) = (\dots, a_{-3}, a_{-2}, \boldsymbol{a_{-1}}, a_0, a_1, \dots),$$

where the boldface indicates the zeroth position.

Theorem 2.1.5. *(i) The bilateral shift is a unitary operator.*

(ii) The adjoint of the bilateral shift, called the backward bilateral shift, *is given by*

$$W^*(\dots, a_{-2}, a_{-1}, \boldsymbol{a_0}, a_1, a_2, \dots) = (\dots, a_{-1}, a_0, \boldsymbol{a_1}, a_2, a_3, \dots).$$

Proof. It is clear that $\|Wx\| = \|x\|$ for all $x \in \ell^2(\mathbb{Z})$, and thus W is an isometry. Define the bounded linear operator A by

$$A(\dots, a_{-2}, a_{-1}, \boldsymbol{a_0}, a_1, a_2, \dots) = (\dots, a_{-1}, a_0, \boldsymbol{a_1}, a_2, a_3, \dots).$$

Obviously, $AW = WA = I$, and thus W is an invertible isometry; i.e., W is a unitary operator.

We need to show that $(Wx, y) = (x, Ay)$ for all x and $y \in \ell^2(\mathbb{Z})$. Let $x = (\dots, a_{-2}, a_{-1}, \boldsymbol{a_0}, a_1, a_2, \dots)$ and $y = (\dots, b_{-2}, b_{-1}, \boldsymbol{b_0}, b_1, b_2, \dots)$. Notice that

$$(Wx, y) = \sum_{n=-\infty}^{\infty} a_{n-1}\overline{b_n}$$

and

$$(x, Ay) = \sum_{n=-\infty}^{\infty} a_n\overline{b_{n+1}}.$$

These sums are equal to each other. Therefore $A = W^*$. □

It will be useful to identify the spectra of the unilateral and bilateral shifts and their adjoints. We first describe the point spectrum of the backward shift.

Theorem 2.1.6. *Let U be the unilateral shift on ℓ^2 and let U^* be its adjoint. Then $\Pi_0(U^*) = \mathbb{D}$. Furthermore, for λ in $\mathbb{D}$, $(U^* - \lambda)f = 0$ for a vector f in ℓ^2 if and only if there exists a constant c such that $f = c(1, \lambda, \lambda^2, \lambda^3, \dots)$.*

Proof. Observe first that, since $\|U^*\| = \|U\| = 1$, the spectral radius formula (Theorem 1.2.4) implies that $\Pi_0(U^*) \subset \sigma(U^*) \subset \overline{\mathbb{D}}$.

If $|\lambda| < 1$, then the vector $f = (1, \lambda, \lambda^2, \lambda^3, \dots)$ is in ℓ^2. Thus

$$U^*(1, \lambda, \lambda^2, \lambda^3, \dots) = (\lambda, \lambda^2, \lambda^3, \lambda^4, \dots) = \lambda\,(1, \lambda, \lambda^2, \lambda^3, \dots)$$

and therefore λ is an eigenvalue for U^*. Hence $\mathbb{D} \subset \Pi_0(U^*)$.

Let $e^{i\theta} \in S^1$. We shall show that $e^{i\theta} \notin \Pi_0(U^*)$. Let $f = (f_0, f_1, f_2, f_3, \dots)$ be a vector in ℓ^2 and suppose that $U^*f = e^{i\theta}f$. This implies

$$(f_1, f_2, f_3, \dots) = (e^{i\theta}f_0, e^{i\theta}f_1, e^{i\theta}f_2, \dots)$$

and therefore that $f_{n+1} = e^{i\theta}f_n$ for all nonnegative integers n. Solving this equation recursively, we obtain $f_n = e^{in\theta}f_0$. Since $f \in \ell^2$, we must have $\{e^{in\theta}f_0\} \to 0$, but this can happen only if $f_0 = 0$. Therefore $f = 0$ and hence $e^{i\theta}$ cannot be an eigenvalue. This shows that $\Pi_0(U^*) = \mathbb{D}$.

To finish the proof we must establish the characterization of the eigenvectors. Let λ be in $\mathbb{D}$ and suppose that $U^*f = \lambda f$ for some nonzero vector f. If $f = (f_0, f_1, f_2, f_3, \dots)$, we have

$$(f_1, f_2, f_3, f_4, \dots) = U^*f = \lambda f = \lambda(f_0, f_1, f_2, f_3, \dots).$$

Thus $f_{n+1} = \lambda f_n$ for all nonnegative integers n. Solving recursively shows that $f_n = \lambda^n f_0$. But this implies that

$$f = f_0(1, \lambda, \lambda^2, \lambda^3, \dots),$$

as desired. □

Theorem 2.1.7. *If U is the unilateral shift on ℓ^2, U^* is its adjoint, W is the bilateral shift on $\ell^2(\mathbb{Z})$, and W^* is its adjoint, then*

(i) $\sigma(U) = \overline{\mathbb{D}}$, $\Pi(U) = S^1$ *and* $\Pi_0(U) = \varnothing$;

(ii) $\sigma(U^*) = \Pi(U^*) = \overline{\mathbb{D}}$ *and* $\Pi_0(U^*) = \mathbb{D}$;

(iii) $\sigma(W) = \Pi(W) = S^1$ *and* $\Pi_0(W) = \varnothing$;

(iv) $\sigma(W^*) = \Pi(W^*) = S^1$ *and* $\Pi_0(W^*) = \varnothing$.

Proof. We shall prove the results for U^* first. Observe that, as seen above (Theorem 2.1.6), $\sigma(U^*) \subset \overline{\mathbb{D}}$ and $\Pi_0(U^*) = \mathbb{D}$. Hence

$$\mathbb{D} = \Pi_0(U^*) \subset \Pi(U^*) \subset \sigma(U^*) \subset \overline{\mathbb{D}}.$$

Since $\sigma(U^*)$ is closed and $S^1 \subset \Pi(U^*)$ (by Theorem 1.2.7), we must have $\overline{\mathbb{D}} = \Pi(U^*) = \sigma(U^*) = \overline{\mathbb{D}}$.

Since $\sigma(U^*) = \overline{\mathbb{D}}$, we have $\sigma(U) = \overline{\mathbb{D}}$ as well. Now, let $\lambda \in \overline{\mathbb{D}}$. We will show that λ is not an eigenvalue of U. Let $f = (f_0, f_1, f_2, f_3, \dots) \in \ell^2$ and suppose that $Uf = \lambda f$. Then

$$(0, f_0, f_1, f_2, \dots) = (\lambda f_0, \lambda f_1, \lambda f_2, \dots).$$

If $\lambda = 0$, this would imply that the left-hand side of the expression above is zero, and thus $f = 0$. If $\lambda \neq 0$, then we can solve the above equation recursively to obtain $f_n = (1/\lambda)^n f_0$ for all n. If we equate the first terms of Uf and λf we obtain $0 = \lambda f_0$ from which we conclude that $f_0 = 0$. Thus $f_n = 0$ for all n and hence λ cannot be an eigenvalue. Therefore $\Pi_0(U) = \varnothing$.

It remains to be shown that $\Pi(U) = S^1$. That will be a consequence of some properties of the spectrum of W.

Since $\|W\| = 1$, the spectral radius formula (Theorem 1.2.4) implies $\sigma(W) \subset \overline{\mathbb{D}}$. On the other hand, since W is invertible and $W^{-1} = W^*$, we have

$$\sigma(W^*) = \left\{ \frac{1}{\lambda} : \lambda \in \sigma(W) \right\} \subset \left\{ \frac{1}{\lambda} : 0 < |\lambda| \leq 1 \right\} = \{\lambda : |\lambda| \geq 1\}.$$

Since $\sigma(W^*) \subset \overline{\mathbb{D}}$ as well, it follows that $\sigma(W^*) \subset S^1$ and hence that $\sigma(W) \subset S^1$.

We claim that $\Pi(U) \subset \Pi(W)$. To see this, let $\lambda \in \Pi(U)$ and let $\{f_n\}$ be a sequence of unit vectors in ℓ^2 such that $\{\|(U - \lambda)f_n\|\}$ approaches 0 as n approaches ∞. Every vector f_n in ℓ^2 corresponds to the vector g_n in $\ell^2(\mathbb{Z})$ whose coordinates in positions $0, 1, 2, \ldots$ are the same as those of f_n and whose coordinates in positions with negative indices are all 0. Clearly, $\|(U - \lambda)f_n\| = \|(W - \lambda)g_n\|$ and $\|g_n\| = \|f_n\| = 1$ for every n. Thus λ is in $\Pi(W)$.

The fact that the boundary of $\sigma(U)$ is contained in $\Pi(U)$ (Theorem 1.2.7) gives

$$S^1 \subset \Pi(U) \subset \Pi(W) \subset \sigma(W) \subset S^1.$$

Therefore $\sigma(W) = \Pi(W) = S^1$ and also $\Pi(U) = S^1$. Since $\sigma(W) = S^1$, we have $\sigma(W^*) = S^1$.

Clearly $\Pi(W^*) \subset S^1$. Let $e^{i\theta} \in S^1$. We will show that $e^{i\theta} \in \Pi(W^*)$. Since $\Pi(W) = S^1$ and $e^{-i\theta} \in S^1$ as well, it follows that there exists a sequence of vectors $\{f_n\}$ of norm 1 in $\ell^2(\mathbb{Z})$ such that $\{\|(W - e^{-i\theta})f_n\|\}$ goes to 0. It is easy to see that, for any vector $f \in \ell^2(\mathbb{Z})$,

$$\|(W - e^{-i\theta})f\| = \|(W^* - e^{i\theta})f\|.$$

Hence $\{\|(W^* - e^{i\theta})f_n\|\}$ goes to 0 as $n \to \infty$, and $e^{i\theta} \in \Pi(W^*)$. Therefore $\Pi(W^*) = S^1$.

Lastly, take $e^{i\theta} \in S^1$. Let

$$x = (\ldots, a_{-2}, a_{-1}, \boldsymbol{a_0}, a_1, a_2, \ldots).$$

If $Wx = e^{i\theta}x$, it follows that $a_{n-1} = e^{i\theta}a_n$ for all integers n. A straightforward induction argument shows that for all integers n we have $a_n = e^{-in\theta}a_0$. Since $x \in \ell^2(\mathbb{Z})$, we must have $\{e^{-in\theta}a_0\} \to 0$ as $n \to \pm\infty$. Hence $a_0 = 0$ and thus x is the zero vector. Since $\Pi_0(W)$ must be contained in S^1, it follows that $\Pi_0(W) = \varnothing$. That $\Pi_0(W^*) = \varnothing$ is proved similarly. □

It is virtually impossible to describe the invariant subspaces of the shift operators in terms of their representations on spaces of sequences. However, complete descriptions of their invariant subspace lattices can be given when they are viewed as operators on $\boldsymbol{H}^2$ and $\boldsymbol{L}^2$. This discovery by Arne Beurling [64] in 1949 led to the modern interest in $\boldsymbol{H}^2$.

Definition 2.1.8. Define the operator M_z ("multiplication by z") on $\boldsymbol{H}^2$ by

$$(M_z f)(z) = z f(z).$$

Clearly, if $f(z) = \sum_{n=0}^{\infty} a_n z^n$, then

$$(M_z f)(z) = \sum_{n=0}^{\infty} a_n z^{n+1}.$$

Therefore M_z acts like the unilateral shift.

Theorem 2.1.9. *The operator M_z on $\boldsymbol{H}^2$ is unitarily equivalent to the unilateral shift.*

Proof. If V is the unitary operator mapping ℓ^2 onto $\boldsymbol{H}^2$ given by

$$V(a_0, a_1, a_2, \dots) = \sum_{n=0}^{\infty} a_n z^n,$$

it is trivial to verify that $VU = M_z V$. □

Thus M_z is a representation of the unilateral shift as an operator on $\boldsymbol{H}^2$; we often refer to M_z as U when no confusion is possible. Notice that $M_z e_k = e_{k+1}$ for $k = 0, 1, 2, \dots$, where $e_k(z) = z^k$.

The bilateral shift has an analogous representation on $\boldsymbol{L}^2$.

Definition 2.1.10. The operators $M_{e^{i\theta}}$ and $M_{e^{-i\theta}}$ are defined on $\boldsymbol{L}^2$ by

$$(M_{e^{i\theta}} f)(e^{i\theta}) = e^{i\theta} f(e^{i\theta}) \quad \text{and} \quad (M_{e^{-i\theta}} f)(e^{i\theta}) = e^{-i\theta} f(e^{i\theta}).$$

Theorem 2.1.11. *The operator $M_{e^{i\theta}}$ on $\boldsymbol{L}^2$ is unitarily equivalent to the bilateral shift W on $\ell^2(\mathbb{Z})$, and the operator $M_{e^{-i\theta}}$ is unitarily equivalent to W^*.*

Proof. If V is the unitary operator mapping $\ell^2(\mathbb{Z})$ onto $\boldsymbol{L}^2$ given by

$$V(\dots, a_{-2}, a_{-1}, \boldsymbol{a_0}, a_1, a_2, \dots) = \sum_{n=-\infty}^{\infty} a_n e^{in\theta},$$

it is easily verified that $VW = M_{e^{i\theta}} V$. Taking adjoints shows that $V^* M_{e^{-i\theta}} = W^* V^*$ and the theorem follows (since V^* is also unitary). □

The following is trivial to verify but important to notice.

Theorem 2.1.12. *The operator $M_{e^{i\theta}}$ leaves the subspace $\widetilde{\boldsymbol{H}}^2$ of $\boldsymbol{L}^2$ invariant and the restriction of $M_{e^{i\theta}}$ to $\widetilde{\boldsymbol{H}}^2$ is the unilateral shift on $\widetilde{\boldsymbol{H}}^2$. On $\ell^2(\mathbb{Z})$, the operator W leaves the subspace ℓ^2, consisting of those sequences whose coordinates in negative positions are 0, invariant, and the restriction of W to ℓ^2 is the unilateral shift on ℓ^2.*

Proof. This is immediate. □

2.2 Invariant and Reducing Subspaces

There are some obvious invariant subspaces of the unilateral shift. Thinking of U as an operator on ℓ^2, it is clear that, for each natural number n, the subspace consisting of those sequences whose first n coordinates are zero is invariant under U. The corresponding invariant subspace for M_z in $\boldsymbol{H}^2$ is the subspace of $\boldsymbol{H}^2$ consisting of the functions whose first n derivatives (including the 0th derivative) vanish at the origin.

The unilateral shift has many invariant subspaces that are very difficult to describe in ℓ^2. We shall see that all the invariant subspaces of the unilateral shift can be explicitly described as subspaces of $\boldsymbol{H}^2$.

One family of such subspaces is the following. For each $z_0 \in \mathbb{D}$ let

$$\mathcal{M}_{z_0} = \left\{ f \in \boldsymbol{H}^2 \ : \ f(z_0) = 0 \right\}.$$

Since $f(z_0) = 0$ implies $z_0 f(z_0) = 0$, it is clear that $\mathcal{M}_{z_0} \in \operatorname{Lat} U$. This can also be obtained as a consequence of the fact that the kernel functions k_{z_0} are eigenvectors for U^*, which implies that $\{k_{z_0}\}^\perp \in \operatorname{Lat} U$ (by Theorem 1.2.20), and

$$\mathcal{M}_{z_0} = \{k_{z_0}\}^\perp.$$

It is easy to determine the reducing subspaces of the unilateral shift.

Theorem 2.2.1. *The only reducing subspaces of the unilateral shift are* $\{0\}$ *and the entire space.*

Proof. This is easily proven using any representation of U. Suppose $\mathcal{M}$ is a subspace of ℓ^2 that reduces U and is different from $\{0\}$. We must show that $\mathcal{M} = \ell^2$.

Since $\mathcal{M} \neq \{0\}$, it follows that there exists a nonzero vector

$$(a_0, a_1, a_2, a_3, \dots) \in \mathcal{M}.$$

Since $\mathcal{M}$ is reducing, it is invariant under both U and U^*. Choose k_0 such that $a_{k_0} \neq 0$. Then $U^{*k_0}(a_0, a_1, a_2, a_3, \dots)$ has its first coordinate different from zero and is in $\mathcal{M}$. By relabeling, we can assume that $a_0 \neq 0$. Then

$$UU^*(a_0, a_1, a_2, a_3, \dots) = (0, a_1, a_2, a_3, \dots),$$

and thus $(0, a_1, a_2, a_3, \dots) \in \mathcal{M}$. It follows that

$$(a_0, a_1, a_2, a_3, \dots) - (0, a_1, a_2, a_3, \dots) = (a_0, 0, 0, 0, \dots) \in \mathcal{M},$$

since $\mathcal{M}$ is a subspace. Dividing by a_0, we see that $(1,0,0,0,\ldots)$ is in $\mathcal{M}$; that is, $e_0 \in \mathcal{M}$. Since $e_n = U^n e_0$ for every n, it follows that $\mathcal{M}$ contains every basis vector e_n, so $\mathcal{M} = \ell^2$. □

The bilateral shift, on the other hand, has many reducing subspaces. To characterize the reducing subspaces of the bilateral shift, it is useful (in light of Theorem 1.2.25) to begin by determining the operators that commute with the bilateral shift.

Definition 2.2.2. The *commutant* of a bounded linear operator A is the set of all bounded linear operators that commute with A.

Definition 2.2.3. Let ϕ be a function in $\boldsymbol{L}^\infty$. The *operator of multiplication by* ϕ, denoted by M_ϕ, is defined by $M_\phi f = \phi f$ for every $f \in \boldsymbol{L}^2$.

Theorem 2.2.4. *If ϕ is a function in $\boldsymbol{L}^\infty$, then $\|M_\phi\| = \|\phi\|_\infty$.*

Proof. Let $f \in \boldsymbol{L}^2$ with $\|f\| = 1$. Since $|\phi(e^{i\theta})| \leq \|\phi\|_\infty$ a.e., it follows that

$$\|M_\phi f\|^2 = \frac{1}{2\pi}\int_0^{2\pi} |\phi(e^{i\theta})f(e^{i\theta})|^2\, d\theta \leq \|\phi\|_\infty^2 \, \frac{1}{2\pi}\int_0^{2\pi} |f(e^{i\theta})|^2\, d\theta.$$

This implies that $\|M_\phi\| \leq \|\phi\|_\infty$.

We now establish the opposite inequality. Let $\lambda_0 = \|\phi\|_\infty$. If $\lambda_0 = 0$ there is nothing to prove, so assume $\lambda_0 \neq 0$. For all natural numbers n, the set

$$E_n = \left\{ e^{i\theta} \, : \, |\phi(e^{i\theta})| > \lambda_0 - \frac{1}{n} \right\}$$

has positive measure. If χ_n is the characteristic function of this set and m is normalized Lebesgue measure on S^1, we have, when n is sufficiently large that $\lambda_0 - 1/n > 0$,

$$\begin{aligned} \|M_\phi \chi_n\|^2 &= \frac{1}{2\pi}\int_{E_n} |\phi(e^{i\theta})|^2\, d\theta \\ &\geq \frac{1}{2\pi}\int_{E_n} \left(\lambda_0 - \frac{1}{n}\right)^2 d\theta \\ &= \left(\lambda_0 - \frac{1}{n}\right)^2 m(E_n). \end{aligned}$$

Also, $\|\chi_n\|^2 = m(E_n)$. It follows that if $f_n = \chi_n / \|\chi_n\|$, then

$$\|M_\phi f_n\| \geq \lambda_0 - \frac{1}{n}$$

for n sufficiently large, and hence

$$\|M_\phi\| \geq \lambda_0 - \frac{1}{n}$$

for n sufficiently large. Thus $\|M_\phi\| \geq \lambda_0 = \|\phi\|_\infty$. □

Theorem 2.2.5. *The commutant of W (regarded as an operator on $\boldsymbol{L}^2$) is*

$$\{M_\phi \,:\, \phi \in \boldsymbol{L}^\infty\}.$$

Proof. Recall that $W = M_{e^{i\theta}}$. If $\phi \in \boldsymbol{L}^\infty$, then clearly M_ϕ commutes with $M_{e^{i\theta}}$ and thus $\{M_\phi : \phi \in \boldsymbol{L}^\infty\}$ is contained in the commutant.

Conversely, assume that A is in the commutant of W. Define $\phi = Ae_0$. Clearly $\phi \in \boldsymbol{L}^2$. We must show that $\phi \in \boldsymbol{L}^\infty$ and that $A = M_\phi$. Since A commutes with W^n for every natural number n, it follows that

$$Ae^{in\theta} = AW^n e_0 = W^n Ae_0 = e^{in\theta} Ae_0 = \phi e^{in\theta}$$

for $n = 0, 1, 2, \ldots$. Since W is invertible, it follows that $AW^{-1} = W^{-1}A$, and thus that $Ae^{in\theta} = \phi e^{in\theta}$ for all integers n. By linearity, it follows that $Ap = \phi\, p$ for all trigonometric polynomials p.

If f is any function in $\boldsymbol{L}^2$, then there exists a sequence of trigonometric polynomials $\{p_n\}$ such that $\{p_n\} \to f$ in $\boldsymbol{L}^2$ as $n \to \infty$. Since A is continuous, it follows that $\{Ap_n\} \to Af$, and thus that $\{\phi p_n\} \to Af$ on $\boldsymbol{L}^2$.

Now, since $\{p_n\} \to f$ in $\boldsymbol{L}^2$, there exists a subsequence, say $\{p_{n_i}\}$, such that $\{p_{n_i}\} \to f$ almost everywhere on S^1. Thus $\{\phi p_{n_i}\} \to \phi f$ almost everywhere. But $\{\phi p_{n_i}\} \to Af$ on $\boldsymbol{L}^2$. Therefore $Af = \phi f$ almost everywhere. That is, $A = M_\phi$.

It remains to be shown that $\phi \in \boldsymbol{L}^\infty$. Fix a natural number n and let $E_n = \{e^{i\theta} \,:\, |\phi(e^{i\theta})| > n\}$. We must show that $m(E_n) = 0$ for n sufficiently large, where m is normalized Lebesgue measure on S^1. Let χ_n be the characteristic function of E_n (which clearly is in $\boldsymbol{L}^2$). Then

$$\|A\chi_n\|^2 = \|\phi\chi_n\|^2 = \frac{1}{2\pi}\int_{E_n} |\phi(e^{i\theta})|^2\, d\theta \geq n^2 m(E_n).$$

Also,

$$\|\chi_n\|^2 = \frac{1}{2\pi}\int_{E_n} d\theta = m(E_n).$$

Thus $\|A\chi_n\|^2 \geq n^2\|\chi_n\|^2$. Therefore if $n > \|A\|$, then $\|\chi_n\| = 0$, so $m(E_n) = 0$. That is, $\phi \in \boldsymbol{L}^\infty$. □

We can now explicitly describe the reducing subspaces of the bilateral shift.

Corollary 2.2.6. *The reducing subspaces of the bilateral shift on $\boldsymbol{L}^2$ are the subspaces*

$$\mathcal{M}_E = \{f \in \boldsymbol{H}^2 : f(e^{i\theta}) = 0 \text{ a.e. on } E\}$$

for measurable subsets $E \subset S^1$.

Proof. Fix any measurable subset E of S^1 and let

$$\mathcal{M}_E = \{f \in \boldsymbol{H}^2 : f(e^{i\theta}) = 0 \text{ a.e. on } E\}.$$

If $f(e^{i\theta_0}) = 0$, then $e^{i\theta_0} f(e^{i\theta_0}) = 0$, so $\mathcal{M}_E$ is invariant under W. Similarly, if $f(e^{i\theta_0}) = 0$, then $e^{-i\theta_0} f(e^{i\theta_0}) = 0$, so $\mathcal{M}_E$ is invariant under W^*. By Theorem 1.2.25, $\mathcal{M}_E$ is reducing.

If $\mathcal{M}$ is a reducing subspace of W and P is the projection onto $\mathcal{M}$, then $WP = PW$, by Theorem 1.2.25. By the previous theorem, $P = M_\phi$ for some $\phi \in \boldsymbol{L}^\infty$. Since P is a projection, $P^2 = P$ and thus $M_\phi^2 = M_\phi$. This implies that $\phi^2 = \phi$ almost everywhere.

But this implies that $\phi = \chi_F$, the characteristic function of the measurable set $F = \{e^{i\theta} \in S^1 : \phi(e^{i\theta}) = 1\}$. Thus $\mathcal{M} = \{f \in \boldsymbol{L}^2 : f\,\chi_F = f\}$. Let E be the complement of F; then $\mathcal{M} = \{f \in \boldsymbol{L}^2 : f(e^{i\theta}) = 0 \text{ a.e. on } E\} = \mathcal{M}_E$. □

A description of the nonreducing invariant subspaces of the bilateral shift can be given.

Theorem 2.2.7. *The subspaces of $\boldsymbol{L}^2$ that are invariant but not reducing for the bilateral shift are of the form $\mathcal{M} = \phi\widetilde{\boldsymbol{H}}^2$, where ϕ is a function in $\boldsymbol{L}^\infty$ such that $|\phi(e^{i\theta})| = 1$ a.e.*

Proof. First note that $\phi \in \boldsymbol{L}^\infty$ and $|\phi(e^{i\theta})| = 1$ a.e. implies that the operator M_ϕ is an isometry, since, for any $f \in \boldsymbol{L}^2$,

$$\|\phi f\|^2 = \frac{1}{2\pi}\int_0^{2\pi} \left|\phi\left(e^{i\theta}\right) f(e^{i\theta})\right|^2 d\theta = \frac{1}{2\pi}\int_0^{2\pi} \left|f(e^{i\theta})\right|^2 d\theta = \|f\|^2.$$

Since M_ϕ is an isometry, $M_\phi\widetilde{\boldsymbol{H}}^2 = \phi\widetilde{\boldsymbol{H}}^2$ is a closed subspace. Since $W\widetilde{\boldsymbol{H}}^2$ is contained in $\widetilde{\boldsymbol{H}}^2$ and M_ϕ commutes with W, it follows that $W\phi\widetilde{\boldsymbol{H}}^2 \subset \phi\widetilde{\boldsymbol{H}}^2$. Hence every subspace of the form $\phi\widetilde{\boldsymbol{H}}^2$ is invariant under W.

It is easily shown that no such subspace reduces W. Given any ϕ as above, $\phi \in \phi\widetilde{\boldsymbol{H}}^2$. But $W^*\phi = e^{-i\theta}\phi \notin \phi\widetilde{\boldsymbol{H}}^2$, since $e^{-i\theta} \notin \widetilde{\boldsymbol{H}}^2$.

Conversely, let $\mathcal{M}$ be any subspace of $\boldsymbol{L}^2$ that is invariant under W but is not reducing. The idea of the proof that $\mathcal{M}$ has the desired form stems from the following:

If $\mathcal{M}$ were equal to $\phi\widetilde{\boldsymbol{H}}^2$, with $|\phi(e^{i\theta})| = 1$ a.e. and $f \in \widetilde{\boldsymbol{H}}^2$, then

$$(\phi e^{i\theta} f, \phi) = \frac{1}{2\pi}\int_0^{2\pi} \overline{\phi(e^{i\theta})}\phi(e^{i\theta})e^{i\theta}f(e^{i\theta})\,d\theta = \frac{1}{2\pi}\int_0^{2\pi} e^{i\theta}f(e^{i\theta})\,d\theta = 0,$$

since the zeroth Fourier coefficient of $e^{i\theta}f$ is zero. Thus ϕ is orthogonal to $e^{i\theta}\phi\widetilde{\boldsymbol{H}}^2$; i.e., $\phi \perp W\mathcal{M}$. Thus if ϕ satisfies the conclusion of the theorem, $\phi \in \mathcal{M} \ominus W\mathcal{M}$. This motivates the choice of ϕ below.

If $W\mathcal{M} = \mathcal{M}$, then $\mathcal{M} = W^{-1}(\mathcal{M}) = W^*(\mathcal{M})$. Thus the assumption that $\mathcal{M}$ is not reducing implies that $W\mathcal{M}$ is a proper subspace of $\mathcal{M}$.

Choose ϕ to be any function in $\mathcal{M} \ominus W\mathcal{M}$ such that $\|\phi\| = 1$. We show that $|\phi(e^{i\theta})| = 1$ a.e. and that $\mathcal{M} = \phi\widetilde{\boldsymbol{H}}^2$.

First of all, since $\phi \perp W\mathcal{M}$, it follows that $\phi \perp W^n\phi$ for all $n \geq 1$. This implies that

$$\frac{1}{2\pi}\int_0^{2\pi} \phi(e^{i\theta})\overline{\phi(e^{i\theta})}e^{-in\theta}\,d\theta = 0 \quad \text{for } n = 1, 2, 3, \ldots,$$

which can be written as

$$\frac{1}{2\pi}\int_0^{2\pi} |\phi(e^{i\theta})|^2 e^{-in\theta}\,d\theta = 0 \quad \text{for } n = 1, 2, 3, \ldots.$$

Taking conjugates, we get

$$\frac{1}{2\pi}\int_0^{2\pi} |\phi(e^{i\theta})|^2 e^{-in\theta}\,d\theta = 0 \quad \text{for } n = \pm 1, \pm 2, \pm 3, \ldots.$$

Thus $|\phi(e^{i\theta})|$ is constant. Since $\|\phi\| = 1$, it follows that $|\phi(e^{i\theta})| = 1$ a.e. Note that this proves, in particular, that $\phi \in \boldsymbol{L}^\infty$.

We now show that $\mathcal{M} = \phi\widetilde{\boldsymbol{H}}^2$. First note that $|\phi(e^{i\theta})| = 1$ a.e. implies that M_ϕ is a unitary operator, since its inverse is $M_{1/\phi}$. Thus M_ϕ sends the orthonormal set $\{e^{in\theta}\}_{n=-\infty}^{\infty}$ to the orthonormal set $\{\phi e^{in\theta}\}_{n=-\infty}^{\infty}$. In particular, it sends the orthonormal basis $\{e^{in\theta}\}_{n\geq 0}$ of $\widetilde{\boldsymbol{H}}^2$ to the orthonormal basis $\{\phi e^{in\theta}\}_{n\geq 0}$ of $\phi\widetilde{\boldsymbol{H}}^2$, and the orthonormal basis $\{e^{in\theta}\}_{n<0}$ of $(\widetilde{\boldsymbol{H}}^2)^\perp$ to the orthonormal basis $\{\phi e^{in\theta}\}_{n<0}$ of $(\phi\widetilde{\boldsymbol{H}}^2)^\perp$.

Since $\phi \in \mathcal{M}$ and $\phi e^{in\theta} = W^n\phi$ for $n \geq 0$, it follows that $\phi\widetilde{\boldsymbol{H}}^2 \subset \mathcal{M}$.

To prove the opposite containment, suppose $f \in \mathcal{M}$. To show that f is in $\phi\widetilde{\boldsymbol{H}}^2$ it suffices to establish that f is orthogonal to $(\phi\widetilde{\boldsymbol{H}}^2)^\perp$.

For $n < 0$,

$$\begin{aligned}(\phi e^{in\theta}, f) &= \frac{1}{2\pi}\int_0^{2\pi} \phi(e^{i\theta})e^{in\theta}\overline{f(e^{i\theta})}\,d\theta \\ &= \frac{1}{2\pi}\int_0^{2\pi} \phi(e^{i\theta})\overline{e^{-in\theta}f(e^{i\theta})}\,d\theta.\end{aligned}$$

Note that, since n is negative, $W^{-n}f \in W\mathcal{M}$. But $W^{-n}f = e^{-in\theta}f$, so $e^{-in\theta}f \in W\mathcal{M}$. Since $\phi \perp W\mathcal{M}$, it follows that

$$\frac{1}{2\pi}\int_0^{2\pi} \phi(e^{i\theta})\overline{e^{-in\theta}f(e^{i\theta})}\,d\theta = 0.$$

Hence $f \perp \left(\phi\widetilde{\boldsymbol{H}}^2\right)^{\perp}$. Or, equivalently, $f \in \phi\widetilde{\boldsymbol{H}}^2$. □

The question arises of the extent to which the invariant subspaces of the bilateral shift uniquely determine the corresponding function ϕ.

Theorem 2.2.8. *If $|\phi_1(e^{i\theta})| = |\phi_2(e^{i\theta})| = 1$, a.e., then $\phi_1\widetilde{\boldsymbol{H}}^2 = \phi_2\widetilde{\boldsymbol{H}}^2$ if and only if there is a constant c of modulus 1 such that $\phi_1 = c\phi_2$.*

Proof. Clearly $\phi_1\widetilde{\boldsymbol{H}}^2 = c\phi_1\widetilde{\boldsymbol{H}}^2$ when $|c| = 1$. Conversely, suppose that $\phi_1\widetilde{\boldsymbol{H}}^2 = \phi_2\widetilde{\boldsymbol{H}}^2$ with $|\phi_1(e^{i\theta})| = |\phi_2(e^{i\theta})| = 1$, a.e. Then there exist functions f_1 and f_2 in $\widetilde{\boldsymbol{H}}^2$ such that

$$\phi_1 = \phi_2 f_2 \quad \text{and} \quad \phi_2 = \phi_1 f_1.$$

Since $|\phi_1(e^{i\theta})| = 1 = |\phi_2(e^{i\theta})|$ a.e., it follows that

$$\phi_1\overline{\phi_2} = f_2 \quad \text{and} \quad \phi_2\overline{\phi_1} = f_1;$$

i.e., $f_1 = \overline{f_2}$. But since f_1 and f_2 are in $\widetilde{\boldsymbol{H}}^2$, $f_1 = \overline{f_2}$ implies that f_1 has Fourier coefficients equal to 0 for all positive and for all negative indices. Since the only nonzero coefficient is in the zeroth place, f_1 and f_2 are constants, obviously having moduli equal to 1. □

Since the unilateral shift is a restriction of the bilateral shift to an invariant subspace, invariant subspaces of the unilateral shift are determined by Theorem 2.2.7: they are the invariant subspaces of the bilateral shift that are contained in $\widetilde{\boldsymbol{H}}^2$. In this case, the functions generating the invariant subspaces are certain analytic functions whose structure is important.

Definition 2.2.9. A function $\phi \in \boldsymbol{H}^\infty$ satisfying $|\widetilde{\phi}(e^{i\theta})| = 1$ a.e. is an *inner function.*

Theorem 2.2.10. *If ϕ is a nonconstant inner function, then $|\phi(z)| < 1$ for all $z \in \mathbb{D}$.*

Proof. This follows immediately from Corollary 1.1.24 and Theorem 1.1.17. □

The definition of inner functions requires that the functions be in $\boldsymbol{H}^\infty$. It is often useful to know that this follows if a function is in $\boldsymbol{H}^2$ and has boundary values of modulus 1 a.e.

Theorem 2.2.11. *Let $\phi \in \boldsymbol{H}^2$. If $|\widetilde{\phi}(e^{i\theta})| = 1$ a.e., then ϕ is an inner function.*

Proof. It only needs to be shown that $\phi \in \boldsymbol{H}^\infty$; this follows from Corollary 1.1.24. □

Corollary 2.2.12 (Beurling's Theorem). *Every invariant subspace of the unilateral shift other than $\{0\}$ has the form $\phi\boldsymbol{H}^2$, where ϕ is an inner function.*

Proof. The unilateral shift is the restriction of multiplication by $e^{i\theta}$ to $\widetilde{\boldsymbol{H}}^2$, so if $\mathcal{M}$ is an invariant subspace of the unilateral shift, it is an invariant subspace of the bilateral shift contained in $\widetilde{\boldsymbol{H}}^2$. Thus, by Theorem 2.2.7, $\mathcal{M} = \phi\widetilde{\boldsymbol{H}}^2$ for some measurable function satisfying $|\phi(e^{i\theta})| = 1$ a.e. (Note that $\{0\}$ is the only reducing subspace of the bilateral shift that is contained in $\widetilde{\boldsymbol{H}}^2$.) Since $1 \in \widetilde{\boldsymbol{H}}^2$, $\phi \in \widetilde{\boldsymbol{H}}^2$.

Translating this situation back to $\boldsymbol{H}^2$ on the disk gives $\mathcal{M} = \phi\boldsymbol{H}^2$ with ϕ inner, by Theorem 2.2.11. □

Corollary 2.2.13. *Every invariant subspace of the unilateral shift is cyclic. (See Definition 1.2.17.)*

Proof. If $\mathcal{M}$ is an invariant subspace of the unilateral shift, it has the form $\phi\boldsymbol{H}^2$ by Beurling's theorem (Corollary 2.2.12). For each n, $U^n\phi = z^n\phi$, so $\bigvee_{n=0}^\infty\{U^n\phi\}$ contains all functions of the form $\phi(z)p(z)$, where p is a polynomial. Since the polynomials are dense in $\boldsymbol{H}^2$ (as the finite sequences are dense in ℓ^2), it follows that $\bigvee_{n=0}^\infty\{U^n\phi\} = \phi\boldsymbol{H}^2$. □

2.3 Inner and Outer Functions

We shall see that every function in $\boldsymbol{H}^2$, other than the constant function 0, can be written as a product of an inner function and a cyclic vector for the unilateral shift. Such cyclic vectors will be shown to have a special form.

Definition 2.3.1. The function $F \in \boldsymbol{H}^2$ is an *outer function* if F is a cyclic vector for the unilateral shift. That is, F is an outer function if

$$\bigvee_{k=0}^{\infty} \{U^k F\} = \boldsymbol{H}^2.$$

Theorem 2.3.2. *If F is an outer function, then F has no zeros in $\mathbb{D}$.*

Proof. If $F(z_0) = 0$, then $(U^n F)(z_0) = z_0^n F(z_0) = 0$ for all n. Since the limit of a sequence of functions in $\boldsymbol{H}^2$ that all vanish at z_0 must also vanish at z_0 (Theorem 1.1.9),

$$\bigvee_{k=0}^{\infty} \{U^k F\}$$

cannot be all of $\boldsymbol{H}^2$. Hence there is no $z_0 \in \mathbb{D}$ with $F(z_0) = 0$. □

Recall that a function analytic on $\mathbb{D}$ is identically zero if it vanishes on a set that has a limit point in $\mathbb{D}$. The next theorem is an analogous result for boundary values of functions in $\boldsymbol{H}^2$.

Theorem 2.3.3 (The F. and M. Riesz Theorem). *If $f \in \boldsymbol{H}^2$ and the set*

$$\left\{e^{i\theta} : \widetilde{f}(e^{i\theta}) = 0\right\}$$

has positive measure, then f is identically 0 *on $\mathbb{D}$.*

Proof. Let $E = \left\{e^{i\theta} \ : \ \widetilde{f}(e^{i\theta}) = 0\right\}$ and let

$$\mathcal{M} = \bigvee_{k=0}^{\infty} \left\{U^k \widetilde{f}\right\} = \bigvee_{k=0}^{\infty} \left\{e^{ik\theta} \widetilde{f}\right\}.$$

Then every function $\widetilde{g} \in \mathcal{M}$ vanishes on E, since all functions $e^{ik\theta}\widetilde{f}$ do. If $\widetilde{f}$ is not identically zero, it follows from Beurling's theorem (Theorem 2.2.12) that $\mathcal{M} = \widetilde{\phi}\widetilde{\boldsymbol{H}^2}$ for some inner function ϕ. In particular, this implies that $\widetilde{\phi} \in \mathcal{M}$, so $\widetilde{\phi}$ vanishes on E. But $|\widetilde{\phi}(e^{i\theta})| = 1$ a.e. This contradicts the hypothesis that E has positive measure, thus $\widetilde{f}$, and hence f, must be identically zero. □

Another beautiful result that follows from Beurling's theorem is the following factorization of functions in $\boldsymbol{H}^2$.

Theorem 2.3.4. *If f is a function in $\boldsymbol{H}^2$ that is not identically zero, then $f = \phi F$, where ϕ is an inner function and F is an outer function. This factorization is unique up to constant factors.*

Proof. Let $f \in \boldsymbol{H}^2$ and consider $\bigvee_{n=0}^{\infty}\{U^n f\}$. If this span is $\boldsymbol{H}^2$, then f is outer by definition, and we can take ϕ to be the constant function 1 and $F = f$ to obtain the desired conclusion.

If $\bigvee_{n=0}^{\infty}\{U^n f\} \neq \boldsymbol{H}^2$, then, by Beurling's theorem (Corollary 2.2.12), there must exist a nonconstant inner function ϕ with $\bigvee_{n=0}^{\infty}\{U^n f\} = \phi\boldsymbol{H}^2$. Since f is in $\bigvee_{n=0}^{\infty}\{U^n f\} = \phi\boldsymbol{H}^2$, there exists a function F in $\boldsymbol{H}^2$ with $f = \phi F$. We shall show that F is outer.

The invariant subspace $\bigvee_{n=0}^{\infty}\{U^n F\}$ equals $\psi\boldsymbol{H}^2$ for some inner function ψ. Then, since $f = \phi F$, it follows that $U^n f = U^n(\phi F) = \phi\, U^n F$ for every positive integer n, from which we can conclude, by taking linear spans, that $\phi\boldsymbol{H}^2 = \phi\psi\boldsymbol{H}^2$. Theorem 2.2.8 now implies that ϕ and $\phi\psi$ are constant multiples of each other. Hence ψ must be a constant function. Therefore $\bigvee_{n=0}^{\infty}\{U^n F\} = \boldsymbol{H}^2$, so F is an outer function.

Note that if $f = \phi F$ with ϕ inner and F outer, then $\bigvee_{n=0}^{\infty}\{U^n f\} = \phi\boldsymbol{H}^2$. Thus uniqueness of the factorization follows from the corresponding assertion in Theorem 2.2.8. □

Definition 2.3.5. For $f \in \boldsymbol{H}^2$, if $f = \phi F$ with ϕ inner and F outer, we call ϕ the *inner part* of f and F the *outer part* of f.

Theorem 2.3.6. *The zeros of an $\boldsymbol{H}^2$ function are precisely the zeros of its inner part.*

Proof. This follows immediately from Theorem 2.3.2 and Theorem 2.3.4. □

To understand the structure of $\operatorname{Lat} U$ as a lattice requires being able to determine when $\phi_1\boldsymbol{H}^2$ is contained in $\phi_2\boldsymbol{H}^2$ for inner functions ϕ_1 and ϕ_2. This will be accomplished by analysis of a factorization of inner functions.

2.4 Blaschke Products

Some of the invariant subspaces of the unilateral shift are those consisting of the functions vanishing at certain subsets of $\mathbb{D}$. The simplest such subspaces are those of the form, for $z_0 \in \mathbb{D}$,

$$\mathcal{M}_{z_0} = \{f \in \boldsymbol{H}^2 \,:\, f(z_0) = 0\}.$$

The subspace $\mathcal{M}_{z_0}$ is an invariant subspace for U. Therefore Beurling's theorem (Corollary 2.2.12) implies that there is an inner function ψ such that $\mathcal{M}_{z_0} = \psi\boldsymbol{H}^2$.

Theorem 2.4.1. *For each $z_0 \in \mathbb{D}$, the function*

$$\psi(z) = \frac{z_0 - z}{1 - \overline{z_0} z}$$

is an inner function and $\mathcal{M}_{z_0} = \{f \in \boldsymbol{H}^2 : f(z_0) = 0\} = \psi \boldsymbol{H}^2$.

Proof. The function ψ is clearly in $\boldsymbol{H}^\infty$. Moreover, it is continuous on the closure of $\mathbb{D}$. Therefore, to show that ψ is inner, it suffices to show that $|\psi(z)| = 1$ when $|z| = 1$. For this, note that $|z| = 1$ implies $z\overline{z} = 1$, so that

$$\left|\frac{z_0 - z}{1 - \overline{z_0} z}\right| = \left|\frac{z_0 - z}{z(\overline{z} - \overline{z_0})}\right| = \frac{1}{|z|}\left|\frac{z_0 - z}{\overline{z} - \overline{z_0}}\right| = 1.$$

To show that $\mathcal{M}_{z_0} = \psi \boldsymbol{H}^2$, first note that $\psi(z_0) f(z_0) = 0$ for all $f \in \boldsymbol{H}^2$, so $\psi \boldsymbol{H}^2 \subset \mathcal{M}_{z_0}$. For the other inclusion, note that $f(z_0) = 0$ implies that $f(z) = \psi(z) g(z)$ for some function g analytic in $\mathbb{D}$.

Let

$$\varepsilon = \inf\left\{|\psi(z)| \, : \, z \in \mathbb{D}, \quad |z| \geq \frac{1 + |z_0|}{2}\right\}.$$

Clearly $\varepsilon > 0$. Thus

$$\frac{1}{2\pi}\int_0^{2\pi} |f(re^{i\theta})|^2 \, d\theta \geq \varepsilon^2 \, \frac{1}{2\pi}\int_0^{2\pi} |g(re^{i\theta})|^2 \, d\theta$$

for $r \geq \frac{1+|z_0|}{2}$. Therefore

$$\sup_{0<r<1} \frac{1}{2\pi}\int_0^{2\pi} |g(re^{i\theta})|^2 \, d\theta \leq \frac{1}{\varepsilon^2} \sup_{0<r<1} \frac{1}{2\pi}\int_0^{2\pi} |f(re^{i\theta})|^2 \, d\theta.$$

It follows from Theorem 1.1.12 that $g \in \boldsymbol{H}^2$. Hence $f = \psi g$ is in $\psi \boldsymbol{H}^2$. □

A similar result holds for subspaces of $\boldsymbol{H}^2$ vanishing on any finite subset of $\mathbb{D}$.

Theorem 2.4.2. *If $z_1, z_2, \ldots, z_n \in \mathbb{D}$,*

$$\mathcal{M} = \left\{f \in \boldsymbol{H}^2 \, : \, f(z_1) = f(z_2) = \cdots = f(z_n) = 0\right\},$$

and

$$\psi(z) = \prod_{k=1}^{n} \frac{z_k - z}{1 - \overline{z_k} z},$$

then ψ is an inner function and $\mathcal{M} = \psi \boldsymbol{H}^2$.

Proof. It is obvious that a product of a finite number of inner functions is inner. Thus Theorem 2.4.1 above implies that ψ is inner.

It is clear that $\psi \boldsymbol{H}^2$ is contained in $\mathcal{M}$. The proof of the opposite inclusion is very similar to the proof of the case of a single factor established in Theorem 2.4.1 above. That is, if $f(z_1) = f(z_2) = \cdots = f(z_n) = 0$, then $f = \psi g$ for some function g analytic on $\mathbb{D}$. It follows as in the previous proof (take r greater than the maximum of $\frac{1+|z_j|}{2}$) that g is in $\boldsymbol{H}^2$, so $f \in \psi \boldsymbol{H}^2$. □

It is important to be able to factor out the zeros of inner functions. If an inner function has only a finite number of zeros in $\mathbb{D}$, such a factorization is implicit in the preceding theorem, as we now show. (We will subsequently consider the case in which an inner function has an infinite number of zeros.) It is customary to distinguish any possible zero at 0.

Corollary 2.4.3. *Suppose that the inner function ϕ has a zero of multiplicity s at 0 and also vanishes at the nonzero points $z_1, z_2, \ldots, z_n \in \mathbb{D}$ (allowing repetition according to multiplicity). Let*

$$\psi(z) = z^s \prod_{k=1}^{n} \frac{z_k - z}{1 - \overline{z_k} z}.$$

Then $\psi(z)$ is an inner function and ϕ can be written as a product $\phi(z) = \psi(z)S(z)$, where S is an inner function.

Proof. Since ψ is a product of inner functions, ψ is inner. The function ϕ is in the subspace $\mathcal{M}$ of the preceding Theorem 2.4.2, so that theorem implies that $\phi = \psi S$, where S is in $\boldsymbol{H}^2$. Moreover, $\widetilde{\phi} = \widetilde{\psi}\widetilde{S}$, so $|\widetilde{S}(e^{i\theta})| = 1$ a.e. Therefore S is an inner function. □

Recall that the Weierstrass factorization theorem asserts that, given any sequence $\{z_j\}$ with $\{|z_j|\} \to \infty$ and any sequence of natural numbers $\{n_j\}$, there exists an entire function whose zeros are precisely the z_j's with multiplicity n_j. It is well known that similar techniques establish that, given any sequence $\{z_j\} \subset \mathbb{D}$ with $\{|z_j|\} \to 1$ as $j \to \infty$ and any sequence of natural numbers $\{n_j\}$, there is a function f analytic on $\mathbb{D}$ whose zeros are precisely the z_j's with multiplicity n_j ([9, p. 169–170], [47, p. 302–303]). For some sequences $\{z_j\}$ there is no such function in $\boldsymbol{H}^2$; it will be important to determine the sequences that can arise as zeros of functions in $\boldsymbol{H}^2$. By Theorem 2.3.6, this reduces to determining the zeros of the inner functions.

There are many sequences $\{z_j\}$ with $\{|z_j|\} \to 1$ that cannot be the set of zeros of a function in $\boldsymbol{H}^2$. To see this, we begin with a fact about products of zeros of inner functions.

Theorem 2.4.4. *If ϕ is an inner function and $\phi(0) \neq 0$, and if $\{z_j\}$ is a sequence in $\mathbb{D}$ such that $\phi(z_j) = 0$ for all j, then $|\phi(0)| < \prod_{j=1}^{n} |z_j|$ for all n.*

Proof. For each natural number n, let

$$B_n(z) = \prod_{j=1}^{n} \frac{z_j - z}{1 - \overline{z_j} z}.$$

As shown in Corollary 2.4.3, each B_n is an inner function and, for each n, there is an inner function S_n such that $\phi = B_n S_n$. By Theorem 2.2.10, $|S_n(z)| < 1$ for all $z \in \mathbb{D}$. Thus $|\phi(z)| < |B_n(z)|$ for $z \in \mathbb{D}$. In particular,

$$|\phi(0)| < |B_n(0)| = \prod_{j=1}^{n} |z_j|.$$

□

Example 2.4.5. *If $z_k = \frac{k}{k+1}$ for natural numbers k, there is no function f in $\boldsymbol{H}^2$ whose set of zeros is exactly $\{z_k\}$.*

Proof. Suppose that f was such a function and let ϕ be its inner part. In particular, $\phi(0) \neq 0$. By the previous theorem,

$$|\phi(0)| < \prod_{j=k}^{n} |z_k|$$

for every natural number n. But

$$\prod_{k=1}^{n} |z_k| = \frac{1}{n+1}.$$

Choosing n large enough so that $\frac{1}{n+1} < |\phi(0)|$ gives a contradiction. □

To describe the zeros of functions in $\boldsymbol{H}^2$ requires some facts about infinite products. We begin with the definition of convergence.

Definition 2.4.6. Given a sequence $\{w_k\}_{k=1}^{\infty}$ of nonzero complex numbers, we say that $\prod_{k=1}^{\infty} w_k$ *converges to P* and write

$$\prod_{k=1}^{\infty} w_k = P$$

if $\{\prod_{k=1}^{n} w_k\} \to P$ as $n \to \infty$ and P is different from 0.

If a finite number of the w_k's are zero, we say that *the product converges to* 0 if there is an N such that $w_k \neq 0$ for $k \geq N$ and

$$\prod_{k=N}^{\infty} w_k$$

converges as defined above.

The restrictions that P be different from 0 and that an infinite product not necessarily be convergent simply because one of its factors is zero are needed in order to insure that convergence of infinite products has properties that we require below.

Corollary 2.4.7. *If $\{z_k\}_{k=1}^{\infty}$ are nonzero zeros of a function f in $\boldsymbol{H}^2$ that is not identically zero, then*

$$\prod_{k=1}^{\infty} |z_k| \qquad \textit{converges.}$$

Proof. If $p_n = \prod_{k=1}^{n} |z_k|$, then $\{p_n\}$ is a decreasing sequence (since $|z_k| < 1$ for all k) and hence converges to some $P \geq 0$. It must be shown that $P > 0$.

If f has a zero of multiplicity m at 0 write $f(z) = z^m g(z)$. Clearly $g \in \boldsymbol{H}^2$. Let ϕ be the inner part of g; the zeros of ϕ are $\{z_k\}_{k=1}^{\infty}$. By Theorem 2.4.4, $\{p_n\}$ is bounded below by $|\phi(0)|$. Therefore $P > 0$ and $\prod_{k=1}^{\infty} |z_k|$ converges. □

Theorem 2.4.8. *If $0 < r_k < 1$ for all k, then $\prod_{k=1}^{\infty} r_k$ converges if and only if $\sum_{k=0}^{\infty}(1 - r_k)$ converges.*

Proof. Assume $\prod_{k=1}^{\infty} r_k$ converges. Since $\{\prod_{k=1}^{n} r_k\}$ converges as $n \to \infty$ to a number different from 0, it follows that

$$\left\{\frac{\prod_{k=1}^{n} r_k}{\prod_{k=1}^{n-1} r_k}\right\} = \{r_n\}$$

converges to 1 as $n \to \infty$. Similarly, if $\sum_{k=0}^{\infty}(1 - r_k)$ converges, then $\{1 - r_n\}$ converges to 0, and thus $\{r_n\}$ converges to 1 as $n \to \infty$. We may, therefore, assume that $\{r_k\} \to 1$.

The product $\prod_{k=1}^{\infty} r_k$ converges if and only if there exists $r > 0$ such that

$$\lim_{n\to\infty} \prod_{k=1}^{n} r_k = r.$$

By continuity of log on $(0, 1]$, this occurs if and only if

$$\lim_{n\to\infty} \log \prod_{k=1}^{n} r_k = \log r,$$

or, equivalently,

$$\lim_{n\to\infty} \sum_{k=1}^{n} \log r_k = \log r.$$

This is the same as convergence of $\sum_{k=1}^{\infty} \log r_k$. Since

$$\lim_{x\to 1^-} \frac{-\log x}{1-x} = 1,$$

the limit comparison test shows that the above series converges if and only if

$$\sum_{k=1}^{\infty} (1 - r_k)$$

converges. □

When a series converges, its "tail" goes to 0. Similarly, when an infinite product converges, its "tail" goes to 1.

Theorem 2.4.9. *Let $0 < r_k < 1$ for all k. If $\prod_{k=1}^{\infty} r_k$ converges, then*

$$\left\{ \prod_{k=m+1}^{n} r_k \right\}$$

converges to 1 as n and m approach infinity.

Proof. Observe that the above sequence is just

$$\frac{\prod_{k=1}^{n} r_k}{\prod_{k=1}^{m} r_k}$$

and hence, when m and n approach infinity, the sequence approaches 1. □

Corollary 2.4.10. *If $\{z_k\}_{k=1}^{\infty}$ are zeros of a function $f \in \boldsymbol{H}^2$ and f is not identically zero, then*

$$\sum_{k=1}^{\infty} (1 - |z_k|) < \infty.$$

Proof. It follows from Corollary 2.4.7 and Theorem 2.4.8 that the subseries obtained by including only the nonzero zeros converges. If the multiplicity of the zero at 0 is m, adding in the terms corresponding to the zeros at 0 adds m to the sum of the subseries. □

Thus if a function in $\boldsymbol{H}^2$ (that is not identically zero) has an infinite number of zeros, the zeros must approach the boundary of $\mathbb{D}$ fairly rapidly. We shall see that the converse also holds; there is an inner function ϕ with zeros at $\{z_k\}_{k=0}^{\infty}$ whenever $\sum_{j=1}^{\infty}(1-|z_j|)$ converges.

Definition 2.4.11. Let $\{z_k\}_{k=1}^{\infty}$ be a sequence of nonzero complex numbers in $\mathbb{D}$ and assume that $\sum_{k=1}^{\infty}(1-|z_k|) < \infty$. Let s be a nonnegative integer. Then the *Blaschke product with zeros* $\{z_k\}$ *and a zero of multiplicity* s *at* 0 is defined by

$$B(z) = z^s \prod_{k=1}^{\infty} \frac{\overline{z_k}}{|z_k|} \frac{z_k - z}{1 - \overline{z_k} z}.$$

Note that s could be zero and there could be only a finite number of (or even no) z_k's.

It should be noted that the $\{z_k\}$ in the definition of a Blaschke product need not be distinct. Coincidence of some of the $\{z_k\}$ is necessary to allow for zeros of multiplicity greater than 1.

We need to show that every Blaschke product converges and has the prescribed zeros. This will be done below; in the course of the proof it will become apparent why the factors $\frac{\overline{z_k}}{|z_k|}$ are required.

We first establish a beautiful classical theorem that will be used in determining the zeros of Blaschke products.

Theorem 2.4.12 (Hurwitz's Theorem). *Let* $\{g_n\}$ *be a sequence of functions that are analytic and have no zeros on a domain* V. *If* $\{g_n\} \to g$ *uniformly on compact subsets of* V, *then either* g *has no zero in* V *or* g *is identically* 0 *on* V.

Proof. Suppose that g is not identically 0 on V. We will show that assuming that g has a zero in V leads to a contradiction. Suppose, then, that $g(z_0) = 0$ for some $z_0 \in V$. Choose $r > 0$ such that z_0 is the only zero of g in the disk $\{z : |z - z_0| \leq r\} \subset V$.

Let $\delta = \min\{\,|g(z)| : |z - z_0| = r\,\}$. Then $\delta > 0$. Since $\{g_n\}$ converges to g uniformly on the compact set $\{z : |z - z_0| = r\}$, it follows that $|g_n(z)| > \delta/2$ for $z \in \{z : |z - z_0| = r\}$ when n is sufficiently large.

Also, $1/g_n$ is analytic in $\{z : |z - z_0| \le r\}$, since $g_n(z) \ne 0$ for all z in $\{z : |z - z_0| \le r\}$. By the above,

$$\left|\frac{1}{g_n(z)}\right| < \frac{2}{\delta} \quad \text{for } z \in \{z : |z - z_0| = r\}.$$

It then follows from the maximum modulus principle ([9, pp. 79, 128], [47, p. 212]) that

$$\left|\frac{1}{g_n(z)}\right| < \frac{2}{\delta} \quad \text{for } z \in \{z : |z - z_0| \le r\},$$

or, equivalently,

$$|g_n(z)| > \frac{\delta}{2} \quad \text{for } z \in \{z : |z - z_0| \le r\}.$$

But this is a contradiction, since $\{g_n(z_0)\} \to g(z_0) = 0$. □

Theorem 2.4.13. *Every Blaschke product*

$$B(z) = z^s \prod_{k=1}^{\infty} \frac{\overline{z_k}}{|z_k|} \frac{z_k - z}{1 - \overline{z_k} z},$$

where s is a nonnegative integer and $\{z_k\}$ is a sequence of nonzero numbers in $\mathbb{D}$ satisfying $\sum_{k=0}^{\infty}(1 - |z_k|) < \infty$, converges for every $z \in \mathbb{D}$. Moreover, B is an inner function whose nonzero zeros are precisely the $\{z_k\}$, counting multiplicity, and a zero of multiplicity s at 0.

Proof. We begin with consideration of only the nonzero zeros. For each natural number n, define the following partial product:

$$B_n(z) = \prod_{k=1}^{n} \frac{\overline{z_k}}{|z_k|} \frac{z_k - z}{1 - \overline{z_k} z}.$$

We must show that $\{B_n\}$ is a Cauchy sequence in $\boldsymbol{H}^2$. Each B_n is a multiple by a constant of modulus one of the corresponding function that was shown to be inner in Theorem 2.4.2, so each B_n is an inner function. Let $n > m$. Then

$$\begin{aligned}
\|B_n - B_m\|^2 &= \frac{1}{2\pi} \int_0^{2\pi} |\widetilde{B_n}(e^{i\theta}) - \widetilde{B_m}(e^{i\theta})|^2 \, d\theta \\
&= \frac{1}{2\pi} \int_0^{2\pi} \left(\widetilde{B_n}(e^{i\theta}) - \widetilde{B_m}(e^{i\theta})\right) \overline{\left(\widetilde{B_n}(e^{i\theta}) - \widetilde{B_m}(e^{i\theta})\right)} \, d\theta \\
&= \frac{1}{2\pi} \int_0^{2\pi} \left(|\widetilde{B_n}(e^{i\theta})|^2 + |\widetilde{B_m}(e^{i\theta})|^2 - 2 \operatorname{Re}\left(\widetilde{B_n}(e^{i\theta})\overline{\widetilde{B_m}(e^{i\theta})}\right)\right) d\theta \\
&= \frac{1}{2\pi} \int_0^{2\pi} \left(2 - 2 \operatorname{Re}\left(\frac{\widetilde{B_n}(e^{i\theta})}{\widetilde{B_m}(e^{i\theta})}\right)\right) d\theta,
\end{aligned}$$

since $|\widetilde{B_n}(e^{i\theta})| = |\widetilde{B_m}(e^{i\theta})| = 1$ a.e. Since $n > m$, $\frac{B_n}{B_m}$ is a finite Blaschke product and hence is, in particular, in $\boldsymbol{H}^2$. Thus

$$\frac{B_n(0)}{B_m(0)} = \frac{1}{2\pi} \int_0^{2\pi} \frac{\widetilde{B_n}(e^{i\theta})}{\widetilde{B_m}(e^{i\theta})} \, d\theta$$

by the Poisson integral formula (Theorem 1.1.21). Therefore

$$\begin{aligned} \|B_n - B_m\|^2 &= 2 - 2 \operatorname{Re} \frac{B_n(0)}{B_m(0)} \\ &= 2 - 2 \operatorname{Re} \prod_{k=m+1}^{n} \frac{\overline{z_k}}{|z_k|} \frac{z_k - 0}{1 - \overline{z_k} 0} \\ &= 2 - 2 \operatorname{Re} \prod_{k=m+1}^{n} \frac{|z_k|^2}{|z_k|} \\ &= 2 - 2 \prod_{k=m+1}^{n} |z_k|. \end{aligned}$$

By Theorem 2.4.8, the infinite product $\prod_{k=1}^{\infty} |z_k|$ converges, and thus by Theorem 2.4.9, $\left\{\prod_{k=m+1}^{n} |z_k|\right\} \to 1$ as n and m approach infinity.

Therefore

$$\|B_n - B_m\|^2 \to 0 \quad \text{as } n, m \to \infty.$$

Thus $\{B_n\}$ is a Cauchy sequence in $\boldsymbol{H}^2$ and therefore it converges in $\boldsymbol{H}^2$ to some function $B_0 \in \boldsymbol{H}^2$.

Convergence in $\boldsymbol{H}^2$ implies uniform convergence on compact subsets of $\mathbb{D}$ (Theorem 1.1.9), so $\{B_n\} \to B_0$ uniformly on compact subsets of $\mathbb{D}$. In particular, the Blaschke product converges at every point $z \in \mathbb{D}$. This implies that $B_0(z_k) = 0$ for all k. To show that

$$B_0(z) = \prod_{k=1}^{\infty} \frac{\overline{z_k}}{|z_k|} \frac{z_k - z}{1 - \overline{z_k} z}$$

as an infinite product, the definition requires that B_0 have no other zeros than the zeros of the factors (i.e., the z_k's). Observe that

$$B_0(0) = \prod_{k=1}^{\infty} |z_k| \neq 0$$

by Corollary 2.4.7, and therefore B_0 is not identically zero.

Suppose that $B_0(z)$ had a zero z_0 different from all the z_k's. Since B_0 is not identically zero there is a closed disk D_0 containing z_0 that is contained in $\mathbb{D}$ and does not contain any of the given zeros. Then $B_0(z_0) = 0$ and B_0 has no other zeros in D_0. Since D_0 is compact, $\{B_n\}$ converges uniformly to B_0 on D_0. This contradicts Hurwitz's theorem (Theorem 2.4.12). Thus B_0 converges as an infinite product.

It is easily seen that B_0 is an inner function, as follows. The convergence of the sequence $\{B_n\}$ to B_0 in $\boldsymbol{H}^2$ implies that $\{\widetilde{B_n}\} \to \widetilde{B_0}$ in $\widetilde{\boldsymbol{H}}^2$, which then implies that there exists a subsequence $\{\widetilde{B_{n_j}}\}$ that converges to $\widetilde{B_0}$ a.e. (since every sequence converging in $\boldsymbol{L}^2$ has a subsequence that converges almost everywhere; see, for example, [47, p. 68]). Since B_n is inner, $|\widetilde{B_n}(e^{i\theta})| = 1$ a.e., so it follows that $|\widetilde{B_0}(e^{i\theta})| = 1$ a.e. Now, $B(z) = z^s B_0(z)$, so B is also an inner function. Since the zeros of B are the zeros of B_0 together with 0, it follows that $B(z)$ has no other zeros than the given ones.

We also want to know that the multiplicity of each zero of B is the same as the number of factors in which it occurs. Clearly, each zero of B has at least that multiplicity. To show that no zero of B_0 has greater multiplicity than the number of factors in which it occurs, proceed as follows. Fix any $\hat{z}$ that occurs as a zero of B_0. Then B_0 can be written as

$$B_0(z) = \left(\prod_{z_k = \hat{z}} \frac{\overline{z_k}}{|z_k|} \frac{z_k - z}{1 - \overline{z_k} z} \right) \left(\prod_{z_k \neq \hat{z}} \frac{\overline{z_k}}{|z_k|} \frac{z_k - z}{1 - \overline{z_k} z} \right).$$

Using Hurwitz' theorem as above, $\displaystyle\prod_{z_k \neq \hat{z}} \frac{\overline{z_k}}{|z_k|} \frac{z_k - z}{1 - \overline{z_k} z}$ does not vanish at $\hat{z}$. Therefore, the multiplicity of $\hat{z}$ as a zero of B_0 is precisely the number of factors with $z_k = \hat{z}$.

The multiplicity of the zero of B at 0 is s since $B_0(0) \neq 0$. □

Corollary 2.4.3 can now be extended to inner functions that have an infinite number of zeros.

Corollary 2.4.14. *Suppose that the inner function ϕ has a zero of multiplicity s at 0 and has nonzero zeros at the points $z_1, z_2, z_3, \ldots$ in $\mathbb{D}$ (repeated according to multiplicity). Let*

$$B(z) = z^s \prod_{k=1}^{\infty} \frac{\overline{z_k}}{|z_k|} \frac{z_k - z}{1 - \overline{z_k} z}$$

be the Blaschke product formed from those zeros. Then ϕ can be written as a product $\phi = BS$, where S is an inner function that has no zeros in $\mathbb{D}$.

Proof. For each positive integer n, let

$$B_n(z) = \prod_{k=1}^{n} \frac{\overline{z_k}}{|z_k|} \frac{z_k - z}{1 - \overline{z_k} z}$$

and

$$B_0(z) = \prod_{k=1}^{\infty} \frac{\overline{z_k}}{|z_k|} \frac{z_k - z}{1 - \overline{z_k} z}.$$

As shown in Theorem 2.4.13, $\{B_n\}$ converges to B_0 uniformly on compact subsets of $\mathbb{D}$, so $\{z^s B_n\}$ converges to B uniformly on compact subsets of $\mathbb{D}$. Then, for each n, the function $\frac{\phi}{z^s B_n}$ is an inner function, since $z^s B_n$ is the product of a constant of modulus one and the function ψ of Corollary 2.4.3.

Therefore, for each $z \in \mathbb{D}$,

$$\left| \frac{\phi(z)}{z^s B_n(z)} \right| < 1.$$

Let $S(z) = \frac{\phi(z)}{B(z)}$. The function S is analytic on $\mathbb{D}$ since, by the previous theorem, the zeros of B are zeros of ϕ with the same multiplicity. Since

$$\left\{ \frac{\phi(z)}{z^s B_n(z)} \right\} \quad \text{converges to } S(z)$$

for every $z \in \mathbb{D}$ except possibly at the zeros of B, it follows that $|S(z)| \leq 1$ except possibly on a countable subset of $\mathbb{D}$. Since an analytic function that has modulus larger than 1 at any point in $\mathbb{D}$ has modulus larger than 1 at an uncountable number of points, this implies that $|S(z)| \leq 1$ for every $z \in \mathbb{D}$. Thus, in particular, $S \in \boldsymbol{H}^2$. Moreover, $\phi = BS$ implies that $|\widetilde{\phi}(e^{i\theta})| = |\widetilde{B}(e^{i\theta})| \, |\widetilde{S}(e^{i\theta})|$ and, since ϕ and B are inner functions, it follows that S is an inner function.

Any zero of S would obviously be a zero of ϕ so, since B has all the zeros of ϕ with the same multiplicity as ϕ does, S has no zeros in $\mathbb{D}$. □

Example 2.4.15. *There is a Blaschke product that is not analytic at any point of* S^1.

Proof. Let $\{c_n\}$ be any sequence dense in S^1 and, for each n, let

$$z_n = \left(1 - \frac{1}{n^2}\right) c_n.$$

Then $1 - |z_n| = \frac{1}{n^2}$, so the $\{z_n\}$ are the zeros of a Blaschke product B.

Moreover, every point of S^1 is a limit point of $\{z_n\}$. If B was analytic at a point on S^1, B would have to vanish at that point and hence be identically zero in a neighborhood of it, which is clearly not the case. □

2.5 The Müntz–Szász Theorem

There is an application of the result that the zeros of an H^∞ function approach the boundary fairly rapidly (Corollary 2.4.10) that is not required in the sequel but that is so beautiful that we cannot resist including it here.

Recall Weierstrass's famous theorem that every function continuous on $[0,1]$ can be uniformly approximated by polynomials [46, p. 159]. In functional-analytic terms, Weierstrass's theorem states that the collection of polynomials is dense in the Banach space $C[0,1]$ of complex-valued functions continuous on $[0,1]$ (with the norm of a function defined by $\|f\| = \sup\{|f(x)| : x \in [0,1]\}$). The following question is very natural: for what subsets S of the set of natural numbers is the linear span of $\{x^n : n \in S\} \cup \{1\}$ dense in $C[0,1]$? (Note that the constant functions must be included, for otherwise every element of the linear span would vanish at zero.)

A remarkable theorem established by Müntz and Szász answers a more general question: it allows nonintegral powers as well.

Theorem 2.5.1 (The Müntz–Szász Theorem). *If $\{p_n\}$ is a sequence of distinct positive numbers that is bounded away from zero, and if $\sum_{n=1}^{\infty} \frac{1}{p_n}$ diverges, then the linear span of the collection $\{x^{p_n}\} \cup \{1\}$ is dense in $C[0,1]$.*

Proof. Recall that the Riesz representation theorem for linear functionals on spaces of continuous functions [47, pp. 129–130] implies that for every bounded linear functional ϕ on $C[0,1]$ there exists a finite regular complex Borel measure μ on $[0,1]$ such that

$$\phi(f) = \int_0^1 f\, d\mu \qquad \text{for all } f \in C[0,1].$$

By the Hahn–Banach theorem, if the given span was not dense in $C[0,1]$ there would exist a bounded linear functional ϕ such that $\phi(1) = \phi(x^{p_n}) = 0$ for all n but ϕ is not the functional 0. For every bounded linear functional that is not identically 0 but satisfies $\phi(1) = 0$, there exists an n such that $\phi(x^n) \neq 0$ (for if $\phi(1) = \phi(x^n) = 0$ for every natural number n, then $\phi(p(x)) = 0$ for every polynomial p, and Weierstrass's theorem implies that ϕ is identically 0).

Therefore, the theorem will be established if it is shown that whenever μ is a finite regular complex Borel measure on $[0,1]$ such that

$$\int_0^1 x^{p_n}\, d\mu(x) = 0 \qquad \text{for all natural numbers } n,$$

it follows that

$$\int_0^1 x^m \, d\mu(x) = 0 \qquad \text{for all natural numbers } m.$$

Suppose, then, that μ is a finite regular complex Borel measure satisfying

$$\int_0^1 x^{p_n} \, d\mu(x) = 0 \qquad \text{for all natural numbers } n.$$

Since $0^{p_n} = 0$,

$$\int_{(0,1]} x^{p_n} \, d\mu(x) = 0 \qquad \text{for all natural numbers } n.$$

Note that if z is a complex number such that $\operatorname{Re} z > 0$, then $|x^z| = |e^{z \log x}| \leq 1$ for $x \in (0,1]$. Thus

$$\int_{(0,1]} x^z \, d\mu(x)$$

exists for every z with $\operatorname{Re} z > 0$. We define the function g on the right half-plane $\{z : \operatorname{Re} z > 0\}$ by

$$g(z) = \int_{(0,1]} x^z \, d\mu(x).$$

Differentiating under the integral sign shows that g is analytic on the right half-plane. Note that $g(p_n) = 0$ for all natural numbers n, and the theorem will be established if we show that $g(m) = 0$ for every natural number m (observe that $\int_{(0,1]} x^m \, d\mu(x) = \int_0^1 x^m \, d\mu(x)$ whether or not μ has an atom at 0). We will show, in fact, that g is identically 0 on $\{z : \operatorname{Re} z > 0\}$.

Recall that the function $z \mapsto \frac{1+z}{1-z}$ is a conformal mapping of $\mathbb{D}$ onto $\{z : \operatorname{Re} z > 0\}$. Define the function f by

$$f(z) = g\left(\frac{1+z}{1-z}\right).$$

Then f is analytic on $\mathbb{D}$. As shown above, $|x^z| \leq 1$ for $\operatorname{Re} z > 0$. Therefore $|g(z)| \leq |\mu|((0,1])$ when $\operatorname{Re} z > 0$, where $|\mu|$ is the total variation of the measure μ. Thus $|f(z)| \leq |\mu|((0,1])$ for $z \in \mathbb{D}$, so f is in $\boldsymbol{H}^\infty$. Since $g(p_n) = 0$ it follows that $f(z_n) = 0$ if $\frac{1+z_n}{1-z_n} = p_n$. This is equivalent to $z_n = \frac{p_n - 1}{p_n + 1}$.

Note that

$$1 - |z_n| = 1 - \left|\frac{p_n - 1}{p_n + 1}\right|.$$

We distinguish two cases. If an infinite number of the p_n's are less than or equal to 1, then, since $\{p_n\}$ is bounded away from zero, $\{p_n\}$ has a limit point in $\{z : \operatorname{Re} z > 0\}$. In this case g is an analytic function vanishing on a set that has a limit point within its domain and therefore is identically zero, which finishes the proof.

In the other case, there exists N such that $p_n > 1$ for $n > N$. For such n,

$$1 - |z_n| = 1 - \left|\frac{p_n - 1}{p_n + 1}\right| = 1 - \frac{p_n - 1}{p_n + 1} = \frac{2}{p_n + 1}.$$

Now the fact that $\sum_{n=N+1}^{\infty} \frac{1}{p_n}$ diverges implies that $\sum_{n=N+1}^{\infty} \frac{2}{p_n+1}$ diverges, as can be seen by the comparison test. Hence $\sum_{n=N+1}^{\infty}(1 - |z_n|)$ diverges and therefore $\sum_{n=1}^{\infty}(1 - |z_n|)$ diverges. Since $f \in \boldsymbol{H}^{\infty}$ and $f(z_n) = 0$ for all n, it follows from Corollary 2.4.10 that $f(z) = 0$ for all $z \in \mathbb{D}$. Therefore $g(z) = 0$ on $\{z : \operatorname{Re} z > 0\}$, finishing the proof. □

There is a converse to this theorem, also due to Müntz and Szasz, that holds for increasing sequences: If $\{p_n\}$ is an increasing sequence of positive numbers such that $\sum_{n=1}^{\infty} \frac{1}{p_n}$ converges, then the closure in $C[0, 1]$ of the linear span of $\{x^{p_n}\} \cup \{1\}$ does not contain x^{α} unless $\alpha = 0$ or $\alpha = p_n$ for some n. For a proof, see [47, p. 313].

One application of the Müntz–Szász theorem is to prime powers. We require a classical result of Euler's.

Theorem 2.5.2 (Euler's Theorem). *If p_j denotes the jth prime number, then the series $\sum_{j=1}^{\infty} \frac{1}{p_j}$ diverges.*

Proof. If the series converged, there would exist an M such that

$$\sum_{j=M}^{\infty} \frac{1}{p_j} < \frac{1}{2}.$$

Assume there exists such an M; we will show that this leads to a contradiction.

Note that the series

$$\sum_{k=0}^{\infty} \left(\sum_{j=M}^{\infty} \frac{1}{p_j} \right)^k$$

would then be a convergent geometric series.

For each $j < M$, the series

$$\sum_{k=0}^{\infty} \frac{1}{p_j^k}$$

is also a convergent geometric series.

Multiplying these series together gives

$$\left(\sum_{k=0}^{\infty}\frac{1}{2^k}\right)\left(\sum_{k=0}^{\infty}\frac{1}{3^k}\right)\left(\sum_{k=0}^{\infty}\frac{1}{5^k}\right)\cdots\left(\sum_{k=0}^{\infty}\frac{1}{p_{M-1}^k}\right)\left(\sum_{k=0}^{\infty}\left(\sum_{j=M}^{\infty}\frac{1}{p_j}\right)^k\right).$$

We will obtain a contradiction by showing that multiplying out the above expression yields a series whose terms contain all the terms of the harmonic series.

To see this we begin as follows. Suppose that the set

$$\{p_{n_1}, p_{n_2}, \ldots, p_{n_s}\}$$

is any collection of prime numbers with $p_{n_i} \geq p_M$ for all i, and suppose that

$$\{\alpha_1, \alpha_2, \ldots, \alpha_s\}$$

are natural numbers. Then

$$\frac{1}{p_{n_1}^{\alpha_1} p_{n_2}^{\alpha_2} \cdots p_{n_s}^{\alpha_s}}$$

is a term that occurs in the expansion of

$$\left(\sum_{j=M}^{\infty}\frac{1}{p_j}\right)^{\alpha_1+\alpha_2+\cdots+\alpha_s}.$$

If n is a natural number greater than 1, it can be written in the form

$$n = p_{m_1}^{\beta_1} p_{m_2}^{\beta_2} \cdots p_{m_t}^{\beta_t} p_{n_1}^{\alpha_1} p_{n_2}^{\alpha_2} \cdots p_{n_s}^{\alpha_s},$$

where each $p_{m_i} < p_M$ and each $p_{n_i} \geq p_M$. Then $\frac{1}{n}$ occurs as a term in the product of

$$\frac{1}{p_{m_1}^{\beta_1}}\frac{1}{p_{m_2}^{\beta_2}}\cdots\frac{1}{p_{m_t}^{\beta_t}}\left(\sum_{j=M}^{\infty}\frac{1}{p_j}\right)^{\alpha_1+\alpha_2+\cdots+\alpha_s}.$$

Thus the expansion of

$$\left(\sum_{k=0}^{\infty}\frac{1}{2^k}\right)\left(\sum_{k=0}^{\infty}\frac{1}{3^k}\right)\left(\sum_{k=0}^{\infty}\frac{1}{5^k}\right)\cdots\left(\sum_{k=0}^{\infty}\frac{1}{p_{M-1}^k}\right)\left(\sum_{k=0}^{\infty}\left(\sum_{j=M}^{\infty}\frac{1}{p_j}\right)^k\right)$$

contains $\frac{1}{n}$ for every natural number n. This contradicts the divergence of the harmonic series. □

Corollary 2.5.3. *Every continuous function on $[0,1]$ is a uniform limit of polynomials whose exponents are prime numbers.*

Proof. Euler's theorem (Theorem 2.5.2) and the Müntz–Szasz theorem (Theorem 2.5.1) immediately imply the corollary. □

2.6 Singular Inner Functions

As we have just seen, every inner function is the product of a Blaschke product and an inner function that has no zeros on $\mathbb{D}$.

Definition 2.6.1. A nonconstant inner function that has no zero in $\mathbb{D}$ is called a *singular inner function.*

The word "singular" is used because of the representation of such functions by singular measures, as will be described below (Theorem 2.6.5).

We begin with an example.

Example 2.6.2. *If*

$$f(z) = \exp\left(\frac{z+1}{z-1}\right),$$

then f is a singular inner function.

Proof. Recall that $|e^w| = |e^{\mathrm{Re}\, w + i \mathrm{Im}\, w}| = |e^{\mathrm{Re}\, w}| = e^{\mathrm{Re}\, w}$ for every complex number w. Hence

$$|f(z)| = \exp\left(\mathrm{Re}\left(\frac{z+1}{z-1}\right)\right).$$

A calculation shows that

$$\mathrm{Re}\left(\frac{z+1}{z-1}\right) = \frac{|z|^2 - 1}{|z-1|^2}.$$

Since this is negative for $z \in \mathbb{D}$, it follows that $|f(z)| < 1$ for all $z \in \mathbb{D}$. Thus $f \in \boldsymbol{H}^\infty$. Moreover, $|z| = 1$ and $z \neq 1$ implies $\mathrm{Re}\, \frac{z+1}{z-1} = 0$, and therefore $|\widetilde{f}(e^{i\theta})| = 1$ for all $\theta \neq 0$. Since e^w is never zero for any complex number w, it follows that f is an inner function with no zeros in $\mathbb{D}$. □

Surprisingly, this simple example suggests the general form of singular inner functions. Slight variants of this example include the functions

$$g(z) = \exp\left(\alpha_0 \frac{z + e^{i\theta_0}}{z - e^{i\theta_0}}\right)$$

for fixed positive numbers α_0 and fixed real numbers θ_0; the same proof as that given above shows that every such g is a singular inner function. Since the product of any finite number of singular inner functions is obviously a singular inner function, any function S of the form

$$S(z) = \exp\left(\alpha_1 \frac{z + e^{i\theta_1}}{z - e^{i\theta_1}} + \alpha_2 \frac{z + e^{i\theta_2}}{z - e^{i\theta_2}} + \cdots + \alpha_n \frac{z + e^{i\theta_n}}{z - e^{i\theta_n}}\right),$$

with the α_j any positive numbers and the θ_j any real numbers, is also a singular inner function.

We will show (Theorem 2.6.5) that every singular inner function is a kind of "continuous infinite product" of functions of this type. A classical representation theorem will be required.

Theorem 2.6.3 (The Herglotz Representation Theorem). *Suppose h is analytic on $\mathbb{D}$ and $\operatorname{Re} h(z) > 0$ for all $z \in \mathbb{D}$. Then there exists a finite positive regular Borel measure μ on $[0, 2\pi]$ such that*

$$h(z) = \frac{1}{2\pi}\int_0^{2\pi} \frac{e^{i\theta}+z}{e^{i\theta}-z}\,d\mu(\theta) + i\,\operatorname{Im} h(0)$$

for $z \in \mathbb{D}$.

Proof. For each $s \in (0,1)$, define h_s by $h_s(z) = h(sz)$ for $z \in \mathbb{D}$. Clearly $h_s \in \boldsymbol{H}^\infty$, and thus $h_s \in \boldsymbol{H}^2$. For each s, the Poisson integral formula (Theorem 1.1.21) yields

$$h_s(re^{it}) = \frac{1}{2\pi}\int_0^{2\pi} \widetilde{h_s}(e^{i\theta}) P_r(\theta - t)\,d\theta$$

for $re^{it} \in \mathbb{D}$. Thus, since $P_r(\theta - t)$ is real-valued,

$$\operatorname{Re}\left(h_s(re^{it})\right) = \frac{1}{2\pi}\int_0^{2\pi} \operatorname{Re}\left(\widetilde{h_s}(e^{i\theta})\right) P_r(\theta - t)\,d\theta.$$

Note that

$$\begin{aligned}
\operatorname{Re}\left(\frac{e^{i\theta}+re^{it}}{e^{i\theta}-re^{it}}\right) &= \operatorname{Re}\left(\frac{e^{i(\theta-t)}+r}{e^{i(\theta-t)}-r}\right) \\
&= \frac{1}{2}\left(\frac{e^{i(\theta-t)}+r}{e^{i(\theta-t)}-r} + \frac{e^{-i(\theta-t)}+r}{e^{-i(\theta-t)}-r}\right) \\
&= \frac{1-r^2}{1-r\left(e^{i(\theta-t)}+e^{-i(\theta-t)}\right)+r^2} \\
&= \frac{1-r^2}{1-2r\cos(\theta-t)+r^2} = P_r(\theta-t).
\end{aligned}$$

If we define a function F by

$$F(z) = \frac{1}{2\pi}\int_0^{2\pi} \frac{e^{i\theta}+z}{e^{i\theta}-z} \operatorname{Re}\left(\widetilde{h_s}(e^{i\theta})\right) d\theta \qquad \text{for } z \in \mathbb{D},$$

then $\operatorname{Re}(F(z)) = \operatorname{Re}(h_s(z))$ for all $z \in \mathbb{D}$. (Note that the integrand is continuous, so the integral is defined even in the Riemann sense.) The function

$$F(z) = \frac{1}{2\pi} \int_0^{2\pi} \frac{e^{i\theta} + z}{e^{i\theta} - z} \operatorname{Re}(\widetilde{h_s}(e^{i\theta}))\, d\theta$$

is analytic in $\mathbb{D}$, as can be seen by simply differentiating under the integral.

Recall that it follows from the Cauchy–Riemann equations that two analytic functions with the same real part differ by at most an imaginary constant [6, p. 63]. Therefore

$$h_s(z) = \frac{1}{2\pi} \int_0^{2\pi} \frac{e^{i\theta} + z}{e^{i\theta} - z} \operatorname{Re}(\widetilde{h_s}(e^{i\theta}))\, d\theta + i \operatorname{Im} h(0),$$

since $\operatorname{Im} F(0) = 0$ and $h_s(0) = h(0)$.

To complete the proof, we must pass from h_s to h itself. Choose a sequence $\{s_n\}$ of positive numbers increasing to 1. For each n, define the positive measure μ_{s_n} on S^1 by

$$\mu_{s_n}(E) = \frac{1}{2\pi} \int_E \operatorname{Re} \widetilde{h_{s_n}}(e^{i\theta})\, d\theta$$

for each Lebesgue-measurable subset E of S^1. The measure μ of the conclusion of the theorem will be obtained as a limit point of $\{\mu_{s_n}\}$. The easiest way to do this is by regarding measures on S^1 as linear functionals on the space of continuous functions on S^1 and using some basic results from the theory of Banach spaces.

Let $C(S^1)$ denote the Banach space of continuous functions on S^1 equipped with the supremum norm. The Riesz representation theorem for bounded linear functionals on such spaces of continuous functions implies that the dual space of $C(S^1)$ is the space of all finite regular complex measures on S^1 with the total variation norm ([12, p. 383], [20, p. 216], [44, p.357], [47, p. 130]). Alaoglu's theorem states that closed balls in a dual space of a Banach space are weak* compact ([12, p. 130], [20, p. 162], [48, p. 68]).

To apply Alaoglu's theorem, it must be shown that $\{\mu_{s_n}\}$ is bounded in norm. Since $\operatorname{Re} \widetilde{h_{s_n}}(e^{i\theta}) \geq 0$ for all θ, each μ_{s_n} is a nonnegative measure on S^1. Thus the total variation norm of μ_{s_n} is simply $\mu_{s_n}(S^1)$. But

$$\mu_{s_n}(S^1) = \frac{1}{2\pi} \int_{S^1} \operatorname{Re} \widetilde{h_{s_n}}(e^{i\theta})\, d\theta = \frac{1}{2\pi} \int_0^{2\pi} \operatorname{Re} \widetilde{h_{s_n}}(e^{i\theta})\, d\theta.$$

It follows from the Cauchy integral formula (Theorem 1.1.19) that

$$h_{s_n}(0) = \frac{1}{2\pi} \int_0^{2\pi} \widetilde{h_{s_n}}(e^{i\theta})\, d\theta.$$

Thus

$$\operatorname{Re} h_{s_n}(0) = \frac{1}{2\pi}\int_0^{2\pi} \operatorname{Re} \widetilde{h_{s_n}}(e^{i\theta})\, d\theta = \mu_{s_n}(S^1).$$

But $h_{s_n}(0) = h(0)$ for all n, so the total variation norm of μ_{s_n} equals $\operatorname{Re} h(0)$ for all n.

Thus the collection $\{\mu_{s_n}\}$ is a bounded subset of the dual space of $C(S^1)$. It is well known that the weak* topology on a closed ball in the dual of a separable Banach space is metrizable ([19, p. 426], [48, p. 70]). Combining this fact with Alaoglu's theorem implies that every bounded sequence in the dual of a separable Banach space has a subsequence that converges in the weak* topology. In particular, then, a subsequence $\{\mu_{s_{n_j}}\}$ converges to some measure μ in the weak* topology. It is easily seen that μ is a nonnegative measure since of all the μ_{s_n}'s are.

Convergence in the weak* topology means that, for every continuous function G on S^1,

$$\left\{\int_0^{2\pi} G(e^{i\theta})\, d\mu_{s_{n_j}}(\theta)\right\}$$

approaches

$$\int_0^{2\pi} G(e^{i\theta})\, d\mu(\theta)$$

as $\{n_j\} \to \infty$. Applying this to the particular function

$$G(e^{i\theta}) = \frac{e^{i\theta} + z}{e^{i\theta} - z},$$

for z any fixed element of $\mathbb{D}$, and recalling that $d\mu_{s_n}(e^{i\theta}) = \frac{1}{2\pi}\operatorname{Re}\widetilde{h_{s_n}}(e^{i\theta})\, d\theta$, shows that

$$\left\{\frac{1}{2\pi}\int_0^{2\pi} \frac{e^{i\theta} + z}{e^{i\theta} - z} \operatorname{Re}\widetilde{h_{s_{n_j}}}(e^{i\theta})\, d\theta + i\operatorname{Im} h(0)\right\}$$

converges to

$$\frac{1}{2\pi}\int_0^{2\pi} \frac{e^{i\theta} + z}{e^{i\theta} - z}\, d\mu(\theta)\, d\theta + i\operatorname{Im} h(0)$$

as $\{n_j\} \to \infty$. But

$$h_{s_{n_j}}(z) = \frac{1}{2\pi}\int_0^{2\pi} \frac{e^{i\theta} + z}{e^{i\theta} - z} \operatorname{Re}\widetilde{h_{s_{n_j}}}(e^{i\theta})\, d\theta + i\operatorname{Im} h(0)$$

for every $z \in \mathbb{D}$. Hence, for $z \in \mathbb{D}$,

$$\left\{h_{s_{n_j}}(z)\right\} \quad \text{converges to} \quad \frac{1}{2\pi}\int_0^{2\pi} \frac{e^{i\theta} + z}{e^{i\theta} - z}\, d\mu(\theta)\, d\theta + i\operatorname{Im} h(0)$$

as $\{n_j\} \to \infty$. Since $h_{s_{n_j}}(z) = h(s_{n_j} z)$ and $\{s_{n_j}\} \to 1$, it follows that $h_{s_{n_j}}(z)$ also converges to $h(z)$ for every $z \in \mathbb{D}$, so

$$h(z) = \frac{1}{2\pi} \int_0^{2\pi} \frac{e^{i\theta} + z}{e^{i\theta} - z} \, d\mu(\theta) + i \mathrm{Im}\, h(0).$$

□

The reason that singular inner functions are called "singular" is because the measures that arise in their representation are singular with respect to Lebesgue measure.

Definition 2.6.4. The complex measure μ defined on all Lebesgue-measurable subsets of S^1 is *singular with respect to Lebesgue measure* m *on* S^1 if there exist disjoint Lebesgue-measurable subsets F and G of S^1 such that $\mu(E) = \mu(E \cap F)$ and $m(E) = m(E \cap G)$ for all measurable subsets E, and $|\mu|(G) = 0$ and $|m|(F) = 0$. (Recall that $|\mu|$ is the total variation of μ; of course, since m is a nonnegative measure, $m = |m|$.)

Theorem 2.6.5. *The singular inner functions (i.e., inner functions that have no zeros on D) have the form*

$$S(z) = K \exp\left(-\frac{1}{2\pi} \int_0^{2\pi} \frac{e^{i\theta} + z}{e^{i\theta} - z} d\mu(\theta) \right)$$

for μ a finite positive regular Borel measure on $[0, 2\pi]$, singular with respect to Lebesgue measure, and K a constant of modulus 1.

Proof. We first show that every function S of the given form is a singular inner function. Let S have the above form. Clearly (differentiate under the integral sign) S is analytic on $\mathbb{D}$. We must show that S is inner and has no zeros in $\mathbb{D}$.

Let $z = re^{it}$ be a point in $\mathbb{D}$. Then

$$\begin{aligned} |S(re^{it})| &= \left| \exp\left(-\frac{1}{2\pi} \int_0^{2\pi} \frac{e^{i\theta} + re^{it}}{e^{i\theta} - re^{it}} \, d\mu(\theta) \right) \right| \\ &= \exp \mathrm{Re} \left(-\frac{1}{2\pi} \int_0^{2\pi} \frac{e^{i\theta} + re^{it}}{e^{i\theta} - re^{it}} \, d\mu(\theta) \right) \\ &= \exp\left(-\frac{1}{2\pi} \int_0^{2\pi} \mathrm{Re} \left(\frac{e^{i\theta} + re^{it}}{e^{i\theta} - re^{it}} \right) \, d\mu(\theta) \right) \\ &= \exp\left(-\frac{1}{2\pi} \int_0^{2\pi} P_r(\theta - t) \, d\mu(\theta) \right). \end{aligned}$$

Since $P_r(\theta - t) > 0$ and μ is a positive measure,

$$-\frac{1}{2\pi}\int_0^{2\pi} P_r(\theta - t)\,d\mu(\theta) < 0.$$

Therefore $|S(re^{it})| < 1$ for all $re^{it} \in \mathbb{D}$.

Since μ is a positive regular Borel measure on S^1, μ is defined by a monotone function α [20, p. 99]. Recall that monotone functions are differentiable almost everywhere. The fact that μ is singular with respect to Lebesgue measure implies that $\alpha'(t) = 0$ a.e. with respect to Lebesgue measure [20, p. 100]. By Fatou's theorem (Theorem 1.1.26),

$$\lim_{r\to 1^-} \frac{1}{2\pi}\int_0^{2\pi} P_r(\theta - t)\,d\mu(\theta) = \alpha'(t) = 0 \quad \text{a.e.}$$

and therefore

$$\lim_{r\to 1^-} |S(re^{it})| = \lim_{r\to 1^-} \exp\left(-\frac{1}{2\pi}\int_0^{2\pi} P_r(\theta - t)\,d\mu(\theta)\right) = 1 \quad \text{a.e.}$$

Thus S is an inner function. Since S is an exponential, S does not have any zeros, and it follows that S is a singular inner function.

Suppose, conversely, that S is a nonconstant inner function with no zeros in $\mathbb{D}$. We must show that there is a singular measure μ such that S has the above form. Since S has no zero in $\mathbb{D}$, we can write $S(z) = \exp(g(z))$ for a function g analytic on $\mathbb{D}$. Since S is an inner function, $|S(z)| < 1$ for $z \in \mathbb{D}$ (Theorem 2.2.10). But $|S(z)| = \exp \operatorname{Re} g(z)$, so $\operatorname{Re} g(z) < 0$ for $z \in \mathbb{D}$. Thus the Herglotz representation theorem (Theorem 2.6.3) applied to the function $(-g)$ shows that there is a finite positive regular Borel measure μ on $[0, 2\pi]$ such that

$$g(z) = -\frac{1}{2\pi}\int_0^{2\pi} \frac{e^{i\theta} + z}{e^{i\theta} - z}\,d\mu(\theta) + i\operatorname{Im} g(0).$$

It follows that

$$S(z) = K \exp\left(-\frac{1}{2\pi}\int_0^{2\pi} \frac{e^{i\theta} + z}{e^{i\theta} - z}\,d\mu(\theta)\right),$$

where $K = \exp(i\operatorname{Im} g(0))$.

It remains to be shown that μ is singular with respect to Lebesgue measure. If α is the monotone function inducing μ, Fatou's theorem (Theorem 1.1.26) gives

$$\lim_{r\to 1^-} |S(re^{it})| = \exp\left(\lim_{r\to 1^-} -\frac{1}{2\pi}\int_0^{2\pi} P_r(\theta - t)\,d\mu(\theta)\right) = \exp(-\alpha'(t)) \quad \text{a.e.}$$

Since $|\widetilde{S}(e^{i\theta})| = 1$ a.e., it follows that $\alpha'(t) = 0$ a.e. Thus the measure μ induced by α is singular with respect to Lebesgue measure. (A measure induced by a function α is singular with respect to Lebesgue measure if and only if $\alpha'(t) = 0$ a.e.; see, for example, [20, p. 100].) $\square$

Corollary 2.6.6. *If ϕ is an inner function, then ϕ can be written as $\phi = BS$, where B is the Blaschke product formed from the zeros of ϕ and S is a singular inner function given by an integral as in Theorem 2.6.5.*

Proof. Given an inner function ϕ, let B be the Blaschke product formed from the zeros of ϕ. By Theorem 2.4.14, ϕ/B is an inner function with no zeros in $\mathbb{D}$. Thus $\phi/B = S$ is a singular inner function and therefore has the form given in Theorem 2.6.5. $\square$

The above analysis of Blaschke products and singular inner functions allows us to completely understand the lattice structure of $\operatorname{Lat} U$. That is, it allows us to precisely describe the invariant subspaces of the unilateral shift that are contained in a given invariant subspace.

Theorem 2.6.7. *Let ϕ_1 and ϕ_2 be inner functions. Then $\phi_1 \boldsymbol{H}^2 \subset \phi_2 \boldsymbol{H}^2$ if and only ϕ_1/ϕ_2 is an inner function. This is equivalent to every zero of ϕ_2 occurring as a zero of ϕ_1 with at least the same multiplicity and, for μ_1 the singular measure corresponding to ϕ_1 and μ_2 the singular measure corresponding to ϕ_2, the inequality $\mu_2(E) \leq \mu_1(E)$ holding for every Borel subset E of $[0, 2\pi]$.*

Proof. First suppose that $\phi_1 = B_1 S_1$ and $\phi_2 = B_2 S_2$ and the zeros of B_2 occur as zeros of B_1 with at least the same multiplicity and that $\mu_1(E) \leq \mu_2(E)$ for every Borel set E, where μ_1 is the singular measure associated with S_1 and μ_2 is the singular measure associated with S_2. Clearly B_1/B_2 is a Blaschke product B_3.

If

$$S_1(z) = K_1 \exp\left(-\frac{1}{2\pi}\int_0^{2\pi} \frac{e^{i\theta} + z}{e^{i\theta} - z}\, d\mu_1(\theta)\right)$$

and

$$S_2(z) = K_2 \exp\left(-\frac{1}{2\pi}\int_0^{2\pi} \frac{e^{i\theta} + z}{e^{i\theta} - z}\, d\mu_2(\theta)\right),$$

then

$$\frac{S_1}{S_2} = \frac{K_1}{K_2} \exp\left\{\left(-\frac{1}{2\pi}\int_0^{2\pi} \frac{e^{i\theta}+z}{e^{i\theta}-z}\, d\mu_1(\theta)\right) - \left(-\frac{1}{2\pi}\int_0^{2\pi} \frac{e^{i\theta}+z}{e^{i\theta}-z}\, d\mu_2(\theta)\right)\right\}$$
$$= \frac{K_1}{K_2} \exp\left(-\frac{1}{2\pi}\int_0^{2\pi} \frac{e^{i\theta}+z}{e^{i\theta}-z}\, d(\mu_1-\mu_2)(\theta)\right).$$

Define the measure μ_3 on S^1 by $\mu_3(E) = \mu_1(E) - \mu_2(E)$ for every Borel set E. Since $\mu_2(E) \le \mu_1(E)$ for all Borel sets E, μ_3 is a nonnegative measure. Moreover, μ_3 is singular with respect to Lebesgue measure since each one of μ_1 and μ_2 is. Hence

$$S_3(z) = \frac{K_1}{K_2} \exp\left(-\frac{1}{2\pi}\int_0^{2\pi} \frac{e^{i\theta}+z}{e^{i\theta}-z}\, d\mu_3(\theta)\right)$$

is a singular inner function.

Since $B_1/B_2 = B_3$ and $S_1/S_2 = S_3$, it follows that

$$\phi_1 = B_1 S_1 = B_2 B_3 S_2 S_3 = \phi_2 B_3 S_3,$$

and hence ϕ_1/ϕ_2 is an inner function. The inner function $B_3 S_3$ is, in particular, in $\boldsymbol{H}^2$, so $\phi_1 \in \phi_2 \boldsymbol{H}^2$. It follows that $\phi_1 \boldsymbol{H}^2 \subset \phi_2 \boldsymbol{H}^2$.

For the converse, let ϕ_1 and ϕ_2 be inner and suppose that $\phi_1 \boldsymbol{H}^2 \subset \phi_2 \boldsymbol{H}^2$. Since $1 \in \boldsymbol{H}^2$, it follows that $\phi_1 \in \phi_2 \boldsymbol{H}^2$ and therefore there exists a function $\phi_3 \in \boldsymbol{H}^2$ such that $\phi_1 = \phi_2 \phi_3$. Taking the boundary values of these $\boldsymbol{H}^2$ functions we get $|\widetilde{\phi_3}(e^{i\theta})| = 1$ a.e. An application of Theorem 2.2.11 shows that ϕ_3 is an inner function; i.e., ϕ_1/ϕ_2 is inner. Write each inner function ϕ_j as a product of the Blaschke product formed by its zeros and a singular inner function given as in the corollary above; i.e., $\phi_j = B_j S_j$. Thus we have

$$B_1 S_1 = B_2 B_3 S_2 S_3.$$

It is clear that we must have $B_1 = B_2 B_3$ and $S_1 = S_2 S_3$ since $B_2 B_3$ is a Blaschke product and $S_2 S_3$ has no zeros.

Since B_2 is a factor of B_1, every zero of B_2 is also a zero of B_1 with at least the same multiplicity. Let μ_i be the measure given by Theorem 2.6.5 for each singular inner function S_i. We then have

$$K_1 \exp\left(-\frac{1}{2\pi}\int_0^{2\pi} \frac{e^{i\theta}+z}{e^{i\theta}-z}\, d\mu_1(\theta)\right)$$
$$= K_2 \exp\left(-\frac{1}{2\pi}\int_0^{2\pi} \frac{e^{i\theta}+z}{e^{i\theta}-z}\, d\mu_2(\theta)\right) K_3 \exp\left(-\frac{1}{2\pi}\int_0^{2\pi} \frac{e^{i\theta}+z}{e^{i\theta}-z}\, d\mu_3(\theta)\right)$$
$$= K_2 K_3 \exp\left(-\frac{1}{2\pi}\int_0^{2\pi} \frac{e^{i\theta}+z}{e^{i\theta}-z}\, d(\mu_2+\mu_3)(\theta)\right)$$

for some constants K_i of modulus 1. Evaluating the above expression at 0 and comparing the polar representation of the resulting complex numbers shows that K_1 must equal K_2K_3. Recall that $\exp(f(z)) = \exp(g(z))$ for functions f and g analytic on a domain implies that there is a real number c such that $f(z) = g(z) + ic$ for all z in the domain. Thus there is a real number c such that

$$\frac{1}{2\pi}\int_0^{2\pi} \frac{e^{i\theta}+z}{e^{i\theta}-z}\, d\mu_1(\theta) = \frac{1}{2\pi}\int_0^{2\pi} \frac{e^{i\theta}+z}{e^{i\theta}-z}\, d(\mu_2+\mu_3)(\theta) + ic$$

for all $z \in \mathbb{D}$. Taking $z = 0$ and recalling that the measures are real-valued shows that $c = 0$. Hence

$$\frac{1}{2\pi}\int_0^{2\pi} \frac{e^{i\theta}+z}{e^{i\theta}-z}\, d\mu_1(\theta) = \frac{1}{2\pi}\int_0^{2\pi} \frac{e^{i\theta}+z}{e^{i\theta}-z}\, d(\mu_2+\mu_3)(\theta).$$

Note that

$$\begin{aligned}
\frac{e^{i\theta}+z}{e^{i\theta}-z} &= \frac{1+ze^{-i\theta}}{1-ze^{-i\theta}} \\
&= (1+ze^{-i\theta})(1+ze^{-i\theta}+z^2e^{-2i\theta}+z^3e^{-3i\theta}+\cdots) \\
&= 1+2\sum_{n=1}^{\infty} z^n e^{-ni\theta}.
\end{aligned}$$

Thus

$$\int_0^{2\pi}\left(1+2\sum_{n=1}^{\infty} z^n e^{-ni\theta}\right) d\mu_1(\theta) = \int_0^{2\pi}\left(1+2\sum_{n=1}^{\infty} z^n e^{-ni\theta}\right) d(\mu_2+\mu_3)(\theta).$$

Therefore

$$\begin{aligned}
&\int_0^{2\pi} d\mu_1(\theta) + 2\sum_{n=1}^{\infty} z^n \int_0^{2\pi} e^{-ni\theta}\, d\mu_1(\theta) \\
&\qquad = \int_0^{2\pi} d(\mu_2+\mu_3)(\theta) + 2\sum_{n=1}^{\infty} z^n \int_0^{2\pi} e^{-ni\theta}\, d(\mu_2+\mu_3)(\theta).
\end{aligned}$$

Each of the power series displayed above converges for $z \in \mathbb{D}$. The power series representation of a function analytic in $\mathbb{D}$ is unique, so

$$\int_0^{2\pi} e^{-ni\theta}\, d\mu_1(\theta) = \int_0^{2\pi} e^{-ni\theta}\, d(\mu_2+\mu_3)(\theta)$$

for each positive integer n. Since each of the μ_j is a nonnegative measure, taking complex conjugates yields

$$\int_0^{2\pi} e^{ni\theta}\, d\mu_1(\theta) = \int_0^{2\pi} e^{ni\theta}\, d(\mu_2+\mu_3)(\theta)$$

for every integer n other than zero. Moreover, considering the constant terms of the above series yields

$$\int_0^{2\pi} d\mu_1(\theta) = \int_0^{2\pi} d(\mu_2+\mu_3)(\theta).$$

Therefore, for all n,

$$\int_0^{2\pi} e^{ni\theta}\, d\mu_1(\theta) = \int_0^{2\pi} e^{ni\theta}\, d(\mu_2+\mu_3)(\theta).$$

Since linear combinations of $\{e^{in\theta}\}$ (i.e., trigonometric polynomials) are uniformly dense in the space of continuous functions on S^1, it follows that $\mu_1 = \mu_2 + \mu_3$. Thus, since the μ_j's are positive measures, $\mu_2(E) \leq \mu_1(E)$ for all Borel sets E. □

It is sometimes useful to know which invariant subspaces of the unilateral shift have finite codimension (i.e., have the property that there is a finite set such that the span of the subspace and that set is all of $\boldsymbol{H}^2$).

Theorem 2.6.8. *If ϕ is an inner function, then $\phi\boldsymbol{H}^2$ has finite codimension if and only if ϕ is a constant multiple of a Blaschke product with a finite number of factors.*

Proof. It is easy to see that the result holds for a Blaschke product with only one factor. If that factor is z, this follows from the obvious fact that $z\boldsymbol{H}^2$ and the constant function 1 span $\boldsymbol{H}^2$. In the case of any other single Blaschke factor, there is a similar situation. For if f_0 is any function in $\boldsymbol{H}^2$ that does not vanish at z_0, the span of f_0 and

$$\left(\frac{\overline{z_0}}{|z_0|}\frac{z_0 - z}{1-\overline{z_o}z}\right)\boldsymbol{H}^2$$

is $\boldsymbol{H}^2$.

The case of an arbitrary finite Blaschke product with an arbitrary finite number of factors follows from a trivial induction. For suppose that it is known that the codimension of $B_n\boldsymbol{H}^2$ is m whenever B_m is a Blaschke product with m factors (repeated according to the multiplicity of the zero). Let B_{m+1} be a Blaschke product with $m+1$ factors, so that $B_{m+1} = \phi_m B_m$, where ϕ_m is a single Blaschke factor and B_m is a product of m Blaschke factors. By the inductive hypothesis, there exists a linearly independent subset $\{f_1, f_2, \ldots, f_m\}$ of $\boldsymbol{H}^2$ such that the span of $B_m\boldsymbol{H}^2$ and $\{f_1, f_2, \ldots, f_m\}$ is $\boldsymbol{H}^2$. It follows that

$$\phi_m\left(B_m\boldsymbol{H}^2 \vee \{f_1, f_2, \ldots, f_m\}\right) = \phi_m\boldsymbol{H}^2,$$

or

$$\phi_m B_m\boldsymbol{H}^2 \vee \{\phi_m f_1, \phi_m f_2, \ldots, \phi_m f_m\} = \phi_m\boldsymbol{H}^2.$$

Choose f_{m+1} such that its span with $\phi_m\boldsymbol{H}^2$ is $\boldsymbol{H}^2$. Then

$$\phi_m B_m\boldsymbol{H}^2 \vee \{\phi_m f_1, \phi_m f_2, \ldots, \phi_m f_m, f_{m+1}\} = \boldsymbol{H}^2.$$

It is easily seen that $\{\phi_m f_1, \phi_m f_2, \ldots, \phi_m f_m, f_{m+1}\}$ is linearly independent.

We must prove the converse. That is, suppose that ϕ is an inner function such that $\phi\boldsymbol{H}^2$ has finite codimension. It must be shown that ϕ is a finite Blaschke product.

By the factorization theorem for inner functions, it suffices to show that ϕ has no singular inner factor and that its Blaschke factor is a finite product. Suppose ϕ had a nonconstant singular part and let $\phi = BS$ where B is a Blaschke product and S is a singular inner function. Let μ be the singular measure determining S; the fact that S is nonconstant means that μ is not identically 0. We define, for each $\alpha \in (0,1)$, a measure μ_α by $\mu_\alpha(E) = \alpha\mu(E)$ for each Borel subset E of $[0, 2\pi]$, and let S_α be the corresponding singular inner function. If $\phi_\alpha = BS_\alpha$ for each α, the preceding theorem (Theorem 2.6.7) implies that the subspaces $\{\phi_\alpha\boldsymbol{H}^2\}$ form an infinite chain of distinct subspaces all of which contain $\phi\boldsymbol{H}^2$. Hence $\phi\boldsymbol{H}^2$ would have infinite codimension. Thus the singular part of ϕ is constant.

All that remains to be shown is that the codimension of $B\boldsymbol{H}^2$ is infinite whenever B is a Blaschke product with an infinite number of factors. If B is a Blaschke product with an infinite number of factors, an infinite chain of subspaces containing $B\boldsymbol{H}^2$ can be obtained by inductively defining B_n as the Blaschke product obtained from B_{n-1} by omitting one of its factors (where B_1 equals B). Then $\{B_n\boldsymbol{H}^2\}$ is such a chain of subspaces (by Theorem 2.6.7).

□

The ordinary definitions of greatest common divisor and least common multiple apply to arbitrary collections of inner functions.

Definition 2.6.9. The inner function ϕ_2 *divides* the inner function ϕ_1 if there exists an inner function ϕ_3 such that $\phi_1 = \phi_2\phi_3$. The inner function ϕ_g is the *greatest common divisor* of the collection $\{\phi_\alpha\}$ of inner functions if ϕ_g divides ϕ_α for every α and ϕ divides ϕ_g whenever ϕ is an inner function that divides all the ϕ_α. The inner function ϕ_m is the *least common multiple* of the collection $\{\phi_\alpha\}$ of inner functions if ϕ_α divides ϕ_m for every α and ϕ_m divides ϕ whenever ϕ is an inner function such that ϕ_α divides ϕ for all α.

Theorem 2.6.10. *Every collection of inner functions has a greatest common divisor. Every finite collection of inner functions has a least common multiple.*

Proof. Let $\{\phi_\alpha\}$ be a collection of inner functions. Define

$$\mathcal{M} = \bigvee \{f : f \in \phi_\alpha \boldsymbol{H}^2 \text{ for some } \alpha\}.$$

It is easily seen that $\mathcal{M}$ is invariant under U. Thus $\mathcal{M} = \phi_g \boldsymbol{H}^2$ for some inner function ϕ_g. Since $\phi_\alpha \boldsymbol{H}^2 \subset \phi_g \boldsymbol{H}^2$ for every α, it follows that ϕ_g divides ϕ_α for all α, by Theorem 2.6.7. Therefore ϕ_g is a common divisor of the ϕ_α. Suppose ϕ is an inner function such that ϕ divides ϕ_α for all α. Then $\phi_\alpha \boldsymbol{H}^2 \subset \phi \boldsymbol{H}^2$ for all α. But then

$$\bigvee \{f : f \in \phi_\alpha \boldsymbol{II}^2 \text{ for some } \alpha\} \subset \phi \boldsymbol{H}^2,$$

and thus $\phi_g \boldsymbol{H}^2 \subset \phi \boldsymbol{H}^2$. This implies that ϕ divides ϕ_g. Therefore ϕ_g is the greatest common divisor. (Notice that ϕ_g could be the constant function 1.)

To prove the second assertion, let $\{\phi_1, \phi_2, \ldots, \phi_n\}$ be a finite collection of inner functions. Let

$$\mathcal{M} = \bigcap_{j=1}^{n} \phi_j \boldsymbol{H}^2.$$

Clearly $(\phi_1 \phi_2 \cdots \phi_n) \in \mathcal{M}$, so $\mathcal{M}$ is not $\{0\}$. Therefore $\mathcal{M} = \phi_l \boldsymbol{H}^2$ for some inner function ϕ_l. Since $\phi_l \boldsymbol{H}^2 \subset \phi_j \boldsymbol{H}^2$, it follows that every ϕ_j divides ϕ_l. Moreover, if ψ is an inner function that is divisible by every ϕ_j, then $\psi \boldsymbol{H}^2 \subset \phi_j \boldsymbol{H}^2$ for each j, so $\psi \boldsymbol{H}^2 \subset \mathcal{M} = \phi_l \boldsymbol{H}^2$. Thus ϕ_l divides ψ. Therefore ϕ_l is the least common multiple. □

Corollary 2.6.11. *If $\mathcal{M}$ is an invariant subspace, other than $\{0\}$, of the unilateral shift, then $\mathcal{M} = \phi \boldsymbol{H}^2$, where ϕ is the greatest common divisor of all the inner parts of all the functions in $\mathcal{M}$.*

Proof. Since $\mathcal{M}$ is an invariant subspace for the unilateral shift, Beurling's theorem (Theorem 2.2.12) guarantees that there exists an inner function ϕ such that $\mathcal{M} = \phi \boldsymbol{H}^2$. We will show that ϕ is the greatest common divisor of all the inner parts of all the functions in $\boldsymbol{H}^2$.

Let $f \in \mathcal{M} = \phi \boldsymbol{H}^2$. Then there exists a function $g \in \boldsymbol{H}^2$ such that $f = \phi g$. Factor the functions f and g into their inner and outer parts: $f = f_I f_O$ and $g = g_I g_O$. We then have $f_I f_O = \phi g_I g_O$, and, by Theorem 2.3.4, there must exist a constant c of modulus 1 such that $f_I = c\phi g_I$. Hence ϕ divides the inner part of f.

Let ψ be an inner function such that ψ divides the inner parts of all functions in $\mathcal{M}$. Since ϕ is inner and is in $\phi \boldsymbol{H}^2 = \mathcal{M}$, it follows that ψ divides ϕ. Therefore ϕ is the greatest common divisor of all the inner parts of all the functions in $\boldsymbol{H}^2$. □

It is of interest to determine what abstract lattices can arise in the form $\operatorname{Lat} A$ for bounded linear operators on a separable Hilbert space.

Definition 2.6.12. The abstract lattice $\mathcal{L}$ is *attainable* if there exists a bounded linear operator A on an infinite-dimensional separable complex Hilbert space such that $\operatorname{Lat} A$ is order-isomorphic to $\mathcal{L}$.

Surprisingly little is known about which lattices are obtainable. The *invariant subspace problem*, the question whether there is an A whose only invariant subspaces are $\{0\}$ and $\mathcal{H}$, can be rephrased as: is the totally ordered lattice with two elements attainable?

For U the unilateral shift, $\operatorname{Lat} U$ is a very complicated and rich lattice, as Theorem 2.6.7 indicates. Some of its sublattices will be of the form $\operatorname{Lat} A$ for suitable operators A.

Recall that given two subspaces $\mathcal{M}$ and $\mathcal{N}$ of a Hilbert space $\mathcal{H}$, the subspace $\mathcal{N} \ominus \mathcal{M}$ is defined to be $\mathcal{N} \cap \mathcal{M}^{\perp}$.

The next theorem shows that an "interval" of an attainable lattice is an attainable lattice.

Theorem 2.6.13. *Let A be a bounded operator on an infinite-dimensional separable Hilbert space. Suppose that $\mathcal{M}$ and $\mathcal{N}$ are in* $\operatorname{Lat} A$ *and $\mathcal{M} \subset \mathcal{N}$. If $\mathcal{N} \ominus \mathcal{M}$ is infinite-dimensional, then the lattice*

$$\{\mathcal{L} : \mathcal{L} \in \operatorname{Lat} A \text{ and } \mathcal{M} \subset \mathcal{L} \subset \mathcal{N}\}$$

is attainable.

Proof. Let P be the projection onto the subspace $\mathcal{N} \ominus \mathcal{M}$. Define the bounded linear operator B on $\mathcal{N} \ominus \mathcal{M}$ as $B = PA\,|_{\mathcal{N} \ominus \mathcal{M}}$. We will show that $\operatorname{Lat} B$ is order-isomorphic to the lattice of the theorem.

Let $\mathcal{K} \in \operatorname{Lat} B$. We show that $\mathcal{M} \oplus \mathcal{K}$ is in the lattice of the theorem. First of all, since $\mathcal{K} \subset \mathcal{N} \ominus \mathcal{M}$, it is clear that $\mathcal{M} \subset \mathcal{M} \oplus \mathcal{K} \subset \mathcal{N}$. Let $m+k \in \mathcal{M} \oplus \mathcal{K}$. Then

$$A(m+k) = Am + (I-P)Ak + PAk.$$

Since $m \in \mathcal{M}$ and $\mathcal{M} \in \operatorname{Lat} A$, it follows that $Am \in \mathcal{M}$. Since $k \in \mathcal{K} \subset \mathcal{N}$ and $\mathcal{N} \in \operatorname{Lat} A$, we have that $Ak \in \mathcal{N}$. Thus, since $I - P$ is the projection onto $\mathcal{N}^\perp \oplus \mathcal{M}$, it follows that $(I-P)Ak \in \mathcal{M}$. Lastly, $PAk = Bk$ and since $k \in \mathcal{K}$ and $\mathcal{K} \in \operatorname{Lat} B$ we must have that $Bk \in \mathcal{K}$. Thus

$$A(m+k) = (Am + (I-P)Ak) + Bk \in \mathcal{M} \oplus \mathcal{K}.$$

Hence $\mathcal{M} \oplus \mathcal{K}$ is a member of the lattice of the theorem.

Now suppose that $\mathcal{L}$ is a member of the given lattice. Define $\mathcal{K} = \mathcal{L} \ominus \mathcal{M}$. Clearly $\mathcal{K} \subset \mathcal{N} \ominus \mathcal{M}$. We will prove that $\mathcal{K} \in \operatorname{Lat} B$. Let $k \in \mathcal{K}$. Since $\mathcal{K} \subset \mathcal{L}$ and $\mathcal{L} \in \operatorname{Lat} A$, we have $Ak \in \mathcal{L}$. Write Ak as $Ak = f + g$ with $f \in \mathcal{L} \ominus \mathcal{M}$ and $g \in \mathcal{M}$. Then $PAk = f$ since P is the projection onto $\mathcal{N} \ominus \mathcal{M}$ and $\mathcal{L} \ominus \mathcal{M} \subset \mathcal{N} \ominus \mathcal{M}$. Thus $Bk = PAk = f \in \mathcal{L} \ominus \mathcal{M} = \mathcal{K}$, so $\mathcal{K} \in \operatorname{Lat} B$. Since $\mathcal{M} \subset \mathcal{L}$, $\mathcal{K} = \mathcal{L} \ominus \mathcal{M}$ is equivalent to $\mathcal{L} = \mathcal{M} \oplus \mathcal{K}$.

Thus $\mathcal{K} \in \operatorname{Lat} B$ if and only if $\mathcal{L} = \mathcal{M} \oplus \mathcal{K}$ is in the lattice in the statement of the theorem, which establishes the isomorphism. □

The invariant subspace lattice of the unilateral shift has interesting "intervals", including the ordinary closed unit interval.

Example 2.6.14. *Let*

$$\phi(z) = \exp\left(\frac{z+1}{z-1}\right)$$

and let $\mathcal{M} = \left(\phi \boldsymbol{H}^2\right)^\perp$. *Then* $\operatorname{Lat}\left(U^* \mid_{\mathcal{M}}\right)$ *is order-isomorphic to the closed unit interval* $[0,1]$ *with its standard ordering.*

Proof. The function ϕ is inner singular. The measure μ defined by $\mu(A) = 2\pi$ for any Borel set A containing 0 and $\mu(B) = 0$ for Borel sets B that do not contain zero is the measure provided by Theorem 2.6.5.

It follows from Beurling's theorem (Theorem 2.2.12) that $\phi \boldsymbol{H}^2 \in \operatorname{Lat} U$. By Theorem 1.2.20, $\mathcal{M} \in \operatorname{Lat} U^*$.

Let $\mathcal{N} \in \operatorname{Lat}\left(U^* \mid_{\mathcal{M}}\right)$. This is equivalent to $\mathcal{M}^\perp \subset \mathcal{N}^\perp$ and $\mathcal{N}^\perp \in \operatorname{Lat} U$. But this means that $\mathcal{N}^\perp = \phi_a \boldsymbol{H}^2$ for some inner function ϕ_a such that ϕ_a divides ϕ. Since ϕ is singular, Theorem 2.6.7 implies that ϕ_a is also singular, and the singular measure μ_a corresponding to ϕ_a must be less than or equal to μ. This is the same as saying that

$$\phi_a(z) = \exp\left(a\frac{z+1}{z-1}\right)$$

for some $a \in [0,1]$. Thus $\text{Lat}\left(U^* \big|_{\mathcal{M}}\right) \cong [0,1]$ as lattices. □

2.7 Outer Functions

The structure of the outer functions can be explicitly described. First we need to establish the following technical lemma.

Lemma 2.7.1. *If $f \in \boldsymbol{H}^2$ and f is not identically* 0*, then* $\log|\widetilde{f}(e^{i\theta})|$ *is in* $\boldsymbol{L}^1(S^1, d\theta)$.

Proof. First of all, write the function log as the difference of its positive and negative parts; that is, $\log x = \log^+ x - \log^- x$.

Let B be the Blaschke product formed from the zeros of f. Then $f = Bg$ for some $g \in \boldsymbol{H}^2$ that never vanishes. Since $|\widetilde{f}(e^{i\theta})| = |\widetilde{g}(e^{i\theta})|$ a.e., it suffices to show that

$$\log|\widetilde{g}(e^{i\theta})| \in \mathbf{L}^1(S^1, d\theta).$$

Since g never vanishes in $\mathbb{D}$, we can write $g(z) = e^{h(z)}$ for some function h analytic in $\mathbb{D}$. Then

$$|g(z)| = \exp(\operatorname{Re} h(z)).$$

Dividing by a constant if necessary, we may assume that $|g(0)| = 1$, or, equivalently, that $\operatorname{Re} h(0) = 0$.

Fix $r \in (0,1)$. Then

$$\frac{1}{2\pi}\int_0^{2\pi} \log|g(re^{i\theta})|\,d\theta = \frac{1}{2\pi}\int_0^{2\pi} \operatorname{Re} h(re^{i\theta})\,d\theta = \operatorname{Re} h(0) = 0,$$

since $\frac{1}{2\pi}\int_0^{2\pi} h(re^{i\theta})\,d\theta = h(0)$ by Cauchy's integral formula (Theorem 1.1.19) applied to the function $h_r(z) = h(rz)$, which is analytic on a neighborhood of $\overline{\mathbb{D}}$.

This implies that

$$\frac{1}{2\pi}\int_0^{2\pi} \log^+|g(re^{i\theta})|\,d\theta = \frac{1}{2\pi}\int_0^{2\pi} \log^-|g(re^{i\theta})|\,d\theta,$$

and thus that

$$\frac{1}{2\pi}\int_0^{2\pi}\left|\log|g(re^{i\theta})|\right|d\theta = 2\,\frac{1}{2\pi}\int_0^{2\pi}\log^+|g(re^{i\theta})|\,d\theta$$
$$= \frac{1}{2\pi}\int_0^{2\pi}\log^+|g(re^{i\theta})|^2\,d\theta.$$

Since $\log^+ x \leq x$ for all $x > 0$ and $g \in \boldsymbol{H}^2$, we then have

$$\frac{1}{2\pi}\int_0^{2\pi}\left|\log|g(re^{i\theta})|\right|d\theta \leq \frac{1}{2\pi}\int_0^{2\pi}|g(re^{i\theta})|^2\,d\theta \leq \|g\|^2.$$

Also, we can choose an increasing sequence $\{r_n\}$ of positive numbers with $\{r_n\} \to 1$ such that $\{g(r_n e^{i\theta})\} \to g(e^{i\theta})$ a.e. (by Corollary 1.1.11). From the above, for each n,

$$\frac{1}{2\pi}\int_0^{2\pi}\left|\log|g(r_n e^{i\theta})|\right|d\theta \leq \|g\|^2.$$

A straightforward application of Fatou's lemma on convergence of Lebesgue integrals [47, p. 23] implies that

$$\frac{1}{2\pi}\int_0^{2\pi}\left|\log|\widetilde{g}(e^{i\theta})|\right|d\theta \leq \|g\|^2,$$

which is the desired result. □

Notice that this gives another proof of the F. and M. Riesz theorem (proven in Theorem 2.3.3 above).

Corollary 2.7.2 (The F. and M. Riesz Theorem). *If $f \in \boldsymbol{H}^2$ and the set*

$$\left\{e^{i\theta} : \widetilde{f}(e^{i\theta}) = 0\right\}$$

has positive measure, then f is identically 0 on $\mathbb{D}$.

Proof. If the set $\{e^{i\theta} : \widetilde{f}(e^{i\theta}) = 0\}$ has positive measure, the function $\log|\widetilde{f}(e^{i\theta})|$ is not integrable on $[0, 2\pi]$. Thus the previous lemma implies that f is identically 0. □

We require a number of well-known inequalities.

Theorem 2.7.3. *(i) For $x > 0$, $\log x \leq x - 1$.*

(ii) Let μ be a measure on a space X with $\mu(X) = 1$ and let g be a positive measurable function. Then

$$\int_X \log g\,d\mu \leq \log \int_X g\,d\mu.$$

(iii) Let μ be a measure on a space X with $\mu(X) = 1$ and let g be a positive measurable function. Then

$$\exp\left(\int_X \log g \, d\mu\right) \leq \int_X g \, d\mu.$$

(iv) If $\log^+$ *and* $\log^-$ *denote the positive and negative parts of the* log *function (such that* $\log x = \log^+ x - \log^- x$*), then*

$$|\log^+ x - \log^+ y| \leq |x - y|$$

for $x, y > 0$.

Proof of (i): If $g(x) = x - 1 - \log x$, then $g(1) = 0$ and $g'(x) = 1 - \frac{1}{x}$ for all $x > 0$. Thus g is decreasing on $(0, 1)$ and increasing on $(1, \infty)$, so $g(x) \geq g(1)$ for all $x > 0$. □

Proof of (ii): Put

$$x = \frac{g}{\int_X g \, d\mu}$$

in the previous inequality and integrate over X. □

Proof of (iii): This follows immediately by exponentiating the previous inequality. □

Proof of (iv): The proof of this inequality is divided into several cases.

Case (a): If $x > 1$ and $y > 1$, then $\log^+ x = \log x$ and $\log^+ y = \log y$. Assume without loss of generality that $y < x$. Applying the mean value theorem yields

$$\frac{|\log x - \log y|}{|x - y|} = \frac{1}{c},$$

for some $c \in (y, x)$. Since $c > 1$, $\frac{1}{c} < 1$, which proves the inequality in this case.

Case (b): If $x \leq 1$ and $y \leq 1$, then $\log^+ x = \log^+ y = 0$, so the inequality is trivial.

Case (c): If $x > 1$ and $y \leq 1$, then we need to show that $\log x \leq x - y$. But the inequality of part *(i)* of this theorem yields $\log x \leq x - 1$. Since $y \leq 1$, it follows that $\log x \leq x - y$. □

We can now return to the investigation of outer functions. We begin by showing that functions of a certain form are outer functions; we subsequently show that every outer function has a representation in that form. We must first establish that functions of the given form are in $\boldsymbol{H}^2$.

Theorem 2.7.4. *If f is in $\boldsymbol{H}^2$ and f is not identically 0 on $\mathbb{D}$, define*

$$F(z) = \exp\left(\frac{1}{2\pi}\int_0^{2\pi} \frac{e^{i\theta}+z}{e^{i\theta}-z} \log|\widetilde{f}(e^{i\theta})|\, d\theta\right).$$

Then F is in $\boldsymbol{H}^2$.

Proof. For each fixed $z \in \mathbb{D}$,

$$\left|\frac{e^{i\theta}+z}{e^{i\theta}-z}\right|$$

is a bounded function of $e^{i\theta} \in S^1$. Since $\log|\widetilde{f}(e^{i\theta})|$ is in $\mathbf{L}^1(S^1, d\theta)$ by Theorem 2.7.1, it follows that $F(z)$ is defined for every $z \in \mathbb{D}$. Clearly, F is analytic on $\mathbb{D}$.

Letting $z = re^{it}$ and recalling that the Poisson kernel is the real part of

$$\frac{e^{i\theta}+re^{it}}{e^{i\theta}-re^{it}}$$

yields

$$|F(re^{it})| = \exp\left(\frac{1}{2\pi}\int_0^{2\pi} P_r(\theta-t)\log|\widetilde{f}(e^{i\theta})|\, d\theta\right).$$

Therefore

$$|F(re^{it})|^2 = \exp\left(\frac{1}{2\pi}\int_0^{2\pi} P_r(\theta-t)\log|\widetilde{f}(e^{i\theta})|^2\, d\theta\right).$$

Applying inequality *(iii)* of Theorem 2.7.3 to this last integral gives

$$|F(re^{it})|^2 \leq \frac{1}{2\pi}\int_0^{2\pi} P_r(\theta-t)|\widetilde{f}(e^{i\theta})|^2\, d\theta.$$

Now integrate with respect to t from 0 to 2π and divide by 2π, getting

$$\frac{1}{2\pi}\int_0^{2\pi} |F(re^{it})|^2\, dt \leq \frac{1}{(2\pi)^2}\int_0^{2\pi}\int_0^{2\pi} P_r(\theta-t)|\widetilde{f}(e^{i\theta})|^2\, d\theta\, dt.$$

Interchanging the order of integration, which is justified by Fubini's theorem [47, p. 164] since $P_r(\theta-t)$ and $|\widetilde{f}(e^{i\theta})|^2$ are nonnegative, we get

$$\frac{1}{2\pi}\int_0^{2\pi} |F(re^{it})|^2\,dt \le \frac{1}{2\pi}\int_0^{2\pi}\left(\frac{1}{2\pi}\int_0^{2\pi} P_r(\theta-t)dt\right)|\widetilde{f}(e^{i\theta})|^2\,d\theta$$
$$= \frac{1}{2\pi}\int_0^{2\pi} |\widetilde{f}(e^{i\theta})|^2\,d\theta = \|f\|^2,$$

since

$$\frac{1}{2\pi}\int_0^{2\pi} P_r(\theta-t)dt = 1 \quad \text{by Corollary 1.1.22.}$$

Taking the supremum over r gives

$$\sup_{0<r<1}\frac{1}{2\pi}\int_0^{2\pi}\left|F(re^{i\theta})\right|^2 d\theta \le \|f\|^2 < \infty,$$

which implies that F is in $\boldsymbol{H}^2$ (by Theorem 1.1.12). □

Corollary 2.7.5. *If f is in $\boldsymbol{H}^2$, f is not identically 0, and F is defined by*

$$F(z) = \exp\left(\frac{1}{2\pi}\int_0^{2\pi}\frac{e^{i\theta}+z}{e^{i\theta}-z}\log|\widetilde{f}(e^{i\theta})|\,d\theta\right),$$

then $|\widetilde{F}(e^{i\theta})| = |\widetilde{f}(e^{i\theta})|$ a.e.

Proof. Since F is in $\boldsymbol{H}^2$,

$$|\widetilde{F}(e^{i\theta})| = \lim_{r\to 1^-}|F(re^{i\theta})| = \exp\left(\lim_{r\to 1^-}\frac{1}{2\pi}\int_0^{2\pi}P_r(\theta-t)\log|\widetilde{f}(e^{i\theta})|\,d\theta\right).$$

By the corollary to Fatou's theorem (Corollary 1.1.27),

$$\exp\left(\lim_{r\to 1^-}\frac{1}{2\pi}\int_0^{2\pi}P_r(\theta-t)\log|\widetilde{f}(e^{i\theta})|\,d\theta\right) = \exp\left(\log|\widetilde{f}(e^{i\theta})|\right) \quad \text{a.e.}$$

Since $\exp\left(\log|\widetilde{f}(e^{i\theta})|\right) = |\widetilde{f}(e^{i\theta})|$, it follows that $|F(e^{i\theta})| = |\widetilde{f}(e^{i\theta})|$ a.e. □

To prove that F is outer, we need the following theorem, which is also very interesting in its own right.

Theorem 2.7.6 (Maximality Property for Outer Functions). *Let f be a function in $\boldsymbol{H}^2$ that is not identically 0. Define the function F by*

$$F(z) = \exp\left(\frac{1}{2\pi}\int_0^{2\pi}\frac{e^{i\theta}+z}{e^{i\theta}-z}\log|\widetilde{f}(e^{i\theta})|\,d\theta\right)$$

for $z\in\mathbb{D}$. If G is any function in $\boldsymbol{H}^2$ satisfying $|\widetilde{G}(e^{i\theta})| = |\widetilde{F}(e^{i\theta})|$, then $|G(z)| \le |F(z)|$ for z in $\mathbb{D}$.

Proof. As usual, by factoring out the Blaschke product we may assume that $G(z) \neq 0$ for all $z \in \mathbb{D}$ and that, therefore, there exists an analytic function h in $\mathbb{D}$ with $G(z) = \exp h(z)$. Since $|G(z)| = \exp(\operatorname{Re} h(z))$, we have $\log|G(z)| = \operatorname{Re} h(z)$.

Let $z = re^{it}$. Since $|\widetilde{G}(e^{i\theta})| = |\widetilde{F}(e^{i\theta})| = |\widetilde{f}(e^{i\theta})|$ a.e. by Corollary 2.7.5 and

$$\log|F(re^{it})| = \frac{1}{2\pi}\int_0^{2\pi} P_r(\theta - t)\log|\widetilde{f}(e^{i\theta})|\,d\theta,$$

the theorem will be established if it is shown that

$$\log|G(re^{it})| \leq \frac{1}{2\pi}\int_0^{2\pi} P_r(\theta - t)\log|\widetilde{G}(e^{i\theta})|\,d\theta$$

for $re^{it} \in \mathbb{D}$.

As we have in a number of previous proofs, we define the function h_s for each $s \in (0,1)$ by $h_s(z) = h(sz)$. Each h_s is in $\boldsymbol{H}^\infty$, so we can write

$$h_s(re^{it}) = \frac{1}{2\pi}\int_0^{2\pi} P_r(\theta - t)h(se^{i\theta})\,d\theta$$

by the Poisson integral formula (Theorem 1.1.21) (since $\widetilde{h_s}(e^{i\theta}) = h(se^{i\theta})$ for all θ).

By taking real parts and recalling that $\log|G(z)| = \operatorname{Re} h(z)$, we obtain

$$\log|G(sre^{it})| = \frac{1}{2\pi}\int_0^{2\pi} P_r(\theta - t)\log|G(se^{i\theta})|\,d\theta.$$

All we need to do now is take limits as s approaches 1 from below. Clearly, $\lim_{s\to 1^-} G(sre^{it}) = G(re^{it})$ for all $re^{it} \in \mathbb{D}$. Therefore

$$\lim_{s\to 1^-} \log|G(sre^{it})| = \log|G(re^{it})|$$

for all $re^{it} \in \mathbb{D}$. Also

$$\begin{aligned}\frac{1}{2\pi}\int_0^{2\pi} P_r(\theta - t)\log|G(se^{i\theta})|\,d\theta = {} & \frac{1}{2\pi}\int_0^{2\pi} P_r(\theta - t)\log^+|G(se^{i\theta})|\,d\theta \\ & - \frac{1}{2\pi}\int_0^{2\pi} P_r(\theta - t)\log^-|G(se^{i\theta})|\,d\theta.\end{aligned}$$

We consider the two integrals separately. First of all,

$$\begin{aligned}
&\left| \frac{1}{2\pi} \int_0^{2\pi} P_r(\theta - t) \log^+ |G(se^{i\theta})| \, d\theta - \frac{1}{2\pi} \int_0^{2\pi} P_r(\theta - t) \log^+ |\widetilde{G}(e^{i\theta})| \, d\theta \right| \\
&= \left| \frac{1}{2\pi} \int_0^{2\pi} P_r(\theta - t) \left(\log^+ |G(se^{i\theta})| - \log^+ |\widetilde{G}(e^{i\theta})| \right) d\theta \right| \\
&\leq \frac{1}{2\pi} \int_0^{2\pi} P_r(\theta - t) \left| \log^+ |G(se^{i\theta})| - \log^+ |\widetilde{G}(e^{i\theta})| \right| d\theta \\
&\leq \frac{1}{2\pi} \int_0^{2\pi} P_r(\theta - t) \left| |G(se^{i\theta})| - |\widetilde{G}(e^{i\theta})| \right| d\theta \quad \text{(by part } (iv) \text{ of Theorem 2.7.3)} \\
&\leq \frac{1}{2\pi} \int_0^{2\pi} P_r(\theta - t) \left| G(se^{i\theta}) - \widetilde{G}(e^{i\theta}) \right| d\theta \\
&\leq \left(\frac{1}{2\pi} \int_0^{2\pi} P_r(\theta - t)^2 \, d\theta \right)^{1/2} \left(\frac{1}{2\pi} \int_0^{2\pi} \left| G(se^{i\theta}) - \widetilde{G}(e^{i\theta}) \right|^2 d\theta \right)^{1/2}
\end{aligned}$$

by the Cauchy–Schwarz inequality. Let

$$M = \left(\frac{1}{2\pi} \int_0^{2\pi} P_r(\theta - t)^2 \, d\theta \right)^{1/2}.$$

It then follows that

$$\begin{aligned}
&\left| \frac{1}{2\pi} \int_0^{2\pi} P_r(\theta - t) \log^+ |G(se^{i\theta})| \, d\theta - \frac{1}{2\pi} \int_0^{2\pi} P_r(\theta - t) \log^+ |\widetilde{G}(e^{i\theta})| \, d\theta \right| \\
&\qquad\qquad \leq M \left(\frac{1}{2\pi} \int_0^{2\pi} \left| G(se^{i\theta}) - \widetilde{G}(e^{i\theta}) \right|^2 d\theta \right)^{1/2}
\end{aligned}$$

for all $s \in (0,1)$.

Therefore, since $G(se^{i\theta})$ converges to $\widetilde{G}(e^{i\theta})$ in $\widetilde{\boldsymbol{H}}^2$ as $s \to 1^-$,

$$\lim_{s \to 1^-} \frac{1}{2\pi} \int_0^{2\pi} P_r(\theta - t) \log^+ |G(se^{i\theta})| \, d\theta = \frac{1}{2\pi} \int_0^{2\pi} P_r(\theta - t) \log^+ |\widetilde{G}(e^{i\theta})| \, d\theta.$$

We now consider

$$\frac{1}{2\pi} \int_0^{2\pi} P_r(\theta - t) \log^- |\widetilde{G}(e^{i\theta})| \, d\theta.$$

Note that

$$\liminf_{s \to 1^-} \log^- |G(se^{i\theta})| = \lim_{s \to 1^-} \log^- |G(se^{i\theta})| = \log^- |\widetilde{G}(e^{i\theta})| \quad \text{a.e.}$$

It follows that

$$\begin{aligned}\frac{1}{2\pi}\int_0^{2\pi} & P_r(\theta-t)\log^-|\widetilde{G}(e^{i\theta})|\,d\theta \\ &= \frac{1}{2\pi}\int_0^{2\pi} P_r(\theta-t)\liminf_{s\to 1^-}\log^-|G(se^{i\theta})|\,d\theta \\ &\le \liminf_{s\to 1^-}\frac{1}{2\pi}\int_0^{2\pi} P_r(\theta-t)\log^-|G(se^{i\theta})|\,d\theta\end{aligned}$$

by Fatou's lemma on convergence of integrals [47, p. 23].

Now write

$$\begin{aligned}\liminf_{s\to 1^-} & \frac{1}{2\pi}\int_0^{2\pi} P_r(\theta-t)\log^-|G(se^{i\theta})|\,d\theta \\ &= \liminf_{s\to 1^-}\frac{1}{2\pi}\int_0^{2\pi} P_r(\theta-t)\left(\log^+|G(se^{i\theta})| - \log|G(se^{i\theta})|\right)d\theta \\ &= \liminf_{s\to 1^-}\left(\frac{1}{2\pi}\int_0^{2\pi} P_r(\theta-t)\log^+|G(se^{i\theta})|\,d\theta\right. \\ &\qquad\qquad \left. - \frac{1}{2\pi}\int_0^{2\pi} P_r(\theta-t)\log|G(se^{i\theta})|\,d\theta\right) \\ &= \liminf_{s\to 1^-}\left(\frac{1}{2\pi}\int_0^{2\pi} P_r(\theta-t)\log^+|G(se^{i\theta})|\,d\theta - \log|G(rse^{it})|\right) \\ &= \lim_{s\to 1^-}\left(\frac{1}{2\pi}\int_0^{2\pi} P_r(\theta-t)\log^+|G(se^{i\theta})|\,d\theta - \log|G(rse^{it})|\right) \\ &= \frac{1}{2\pi}\int_0^{2\pi} P_r(\theta-t)\log^+|\widetilde{G}(e^{i\theta})|\,d\theta - \log|G(re^{it})|,\end{aligned}$$

since, as was established above,

$$\log|G(sre^{it})| = \frac{1}{2\pi}\int_0^{2\pi} P_r(\theta-t)\log|G(se^{i\theta})|\,d\theta.$$

Combining the last two calculations results in

$$\begin{aligned}\log|G(re^{it})| \le & \frac{1}{2\pi}\int_0^{2\pi} P_r(\theta-t)\log^+|\widetilde{G}(e^{i\theta})|\,d\theta \\ & - \frac{1}{2\pi}\int_0^{2\pi} P_r(\theta-t)\log^-|\widetilde{G}(e^{i\theta})|\,d\theta\end{aligned}$$

and therefore

$$\log|G(re^{it})| \le \frac{1}{2\pi}\int_0^{2\pi} P_r(\theta-t)\log|G(e^{i\theta})|\,d\theta,$$

as desired. □

Now we can finally prove that functions of the above form are outer.

Theorem 2.7.7. *If f is in $\boldsymbol{H}^2$ and F is defined by*

$$F(z) = \exp\left(\frac{1}{2\pi}\int_0^{2\pi} \frac{e^{i\theta}+z}{e^{i\theta}-z}\log|f(e^{i\theta})|\,d\theta\right),$$

then F is outer.

Proof. It was shown above (Theorem 2.7.4) that F is in $\boldsymbol{H}^2$. Thus $F = \phi G$ for ϕ an inner function and G an outer function (Theorem 2.3.4). It suffices to show that ϕ is a constant function. Since $|\widetilde{\phi}(e^{i\theta})| = 1$ a.e., it follows that $|\widetilde{F}(e^{i\theta})| = |\widetilde{G}(e^{i\theta})|$ a.e.

If ϕ was not a constant function, we would have $|\phi(z)| < 1$ for all $z \in \mathbb{D}$ (by Theorem 2.2.10). Thus $|F(z)| = |\phi(z)|\,|G(z)| < |G(z)|$ for all $z \in \mathbb{D}$. This contradicts the maximality property of F established in the above theorem (Theorem 2.7.6). □

There is a complete, explicit, classification of outer functions.

Corollary 2.7.8. *The function G in $\boldsymbol{H}^2$ is outer if and only if there exists a constant K of modulus 1 such that*

$$G(z) = K\exp\left(\frac{1}{2\pi}\int_0^{2\pi} \frac{e^{i\theta}+z}{e^{i\theta}-z}\log|\widetilde{G}(e^{i\theta})|\,d\theta\right)$$

for all $z \in \mathbb{D}$.

Proof. Since a nonzero constant times an outer function is outer, the previous theorem establishes that every function of the given form is outer.

Conversely, suppose G is any outer function and define

$$F(z) = \exp\left(\frac{1}{2\pi}\int_0^{2\pi} \frac{e^{i\theta}+z}{e^{i\theta}-z}\log|\widetilde{G}(e^{i\theta})|\,d\theta\right).$$

By Corollary 2.7.5, it follows that $|\widetilde{F}(e^{i\theta})| = |\widetilde{G}(e^{i\theta})|$ a.e. Therefore Theorem 2.7.6 gives $|G(z)| \leq |F(z)|$ for all $z \in \mathbb{D}$. Define the function ϕ by

$$\phi(z) = \frac{G(z)}{F(z)}$$

for all $z \in \mathbb{D}$. This function is clearly analytic in $\mathbb{D}$ and, since $|G(z)| \leq |F(z)|$, we get $|\phi(z)| \leq 1$ for all $z \in \mathbb{D}$. Hence ϕ is in $\boldsymbol{H}^\infty$. Note that

$$|\widetilde{\phi}(e^{i\theta})| = \lim_{r\to 1^-}|\phi(re^{i\theta})| = \lim_{r\to 1^-}\frac{|G(re^{i\theta})|}{|F(re^{i\theta})|} = 1 \quad \text{a.e.}$$

Thus ϕ is inner. Since $G = \phi F$ and both G and F are outer, ϕ must be some constant K (by Theorem 2.3.4). □

Our knowledge of inner and outer functions produces a precise description of all the functions in $\boldsymbol{H}^2$.

Corollary 2.7.9. *Let f be a function in $\boldsymbol{H}^2$ that is not identically 0, and let B denote the Blaschke product formed from its zeros. Then there exists a constant K of modulus 1 and a singular measure μ on S^1 such that $f(z)$ has the form*

$$KB(z)\exp\left(-\frac{1}{2\pi}\int_0^{2\pi}\frac{e^{i\theta}+z}{e^{i\theta}-z}\,d\mu(\theta)\right)\exp\left(\frac{1}{2\pi}\int_0^{2\pi}\frac{e^{i\theta}+z}{e^{i\theta}-z}\log|\widetilde{f}(e^{i\theta})|\,d\theta\right)$$

for all $z\in\mathbb{D}$.

Proof. This follows immediately from Theorem 2.3.4, Corollary 2.4.14, Theorem 2.6.5, and Corollary 2.7.8. □

There is another nice characterization of outer functions.

Theorem 2.7.10. *The function F in $\boldsymbol{H}^2$ is outer if and only if*

$$\log|F(0)| = \frac{1}{2\pi}\int_0^{2\pi}\log|\widetilde{F}(e^{i\theta})|\,d\theta.$$

Proof. If $F\in\boldsymbol{H}^2$ is outer, then

$$F(z) = K\exp\left(\frac{1}{2\pi}\int_0^{2\pi}\frac{e^{i\theta}+z}{e^{i\theta}-z}\log|\widetilde{F}(e^{i\theta})|\,d\theta\right)$$

for some K (Corollary 2.7.8).

Then,

$$F(0) = K\exp\left(\frac{1}{2\pi}\int_0^{2\pi}\log|\widetilde{F}(e^{i\theta})|\,d\theta\right),$$

so

$$\log|F(0)| = \frac{1}{2\pi}\int_0^{2\pi}\log|\widetilde{F}(e^{i\theta})|\,d\theta.$$

To prove the converse, suppose that F is an $\boldsymbol{H}^2$ function satisfying

$$\log|F(0)| = \frac{1}{2\pi}\int_0^{2\pi}\log|\widetilde{F}(e^{i\theta})|\,d\theta.$$

The function F can be factored in the form $F=\phi G$ with ϕ an inner function and G an outer function (Theorem 2.3.4). Since G is outer, the first part of this proof shows that

$$\log|G(0)| = \frac{1}{2\pi}\int_0^{2\pi}\log|\widetilde{G}(e^{i\theta})|\,d\theta.$$

However, since ϕ is inner, $|\widetilde{F}(e^{i\theta})| = |\widetilde{G}(e^{i\theta})|$ a.e. Therefore

$$\log|G(0)| = \frac{1}{2\pi}\int_0^{2\pi} \log|\widetilde{F}(e^{i\theta})|\, d\theta.$$

By hypothesis, the last expression equals $\log|F(0)|$. Therefore $|F(0)| = |G(0)|$, which means that $|\phi(0)| = 1$. But since ϕ is an inner function, the maximum modulus theorem ([9, pp. 79, 128], [47, p. 212]) implies that ϕ is a constant of modulus 1, so F is a nonzero multiple of G and is therefore outer. □

2.8 Exercises

2.1. Let $\mathcal{M}_E = \{f \in \boldsymbol{L}^2 : f(e^{i\theta}) = 0, \text{ a.e. on } E\}$ for a measurable set E whose (normalized) Lebesgue measure is neither 0 nor 1. Verify that $\mathcal{M}_E$ is a nontrivial (closed) subspace of $\boldsymbol{L}^2$.

2.2. If U is the unilateral shift and A is an operator such that $\operatorname{Lat} U \subset \operatorname{Lat} A$, show that $AU = UA$. (Hint: Consider adjoints.)

2.3. Prove that the numerical range of the unilateral shift is the open unit disk. (Hint: One approach uses Exercise 1.12 in Chapter 1.)

2.4. Suppose that ϕ is in $\widetilde{\boldsymbol{H}}^\infty$.

(i) Show that M_ϕ is an isometry on $\widetilde{\boldsymbol{H}}^2$ (i.e., $\|M_\phi f\| = \|f\|$ for all $f \in \widetilde{\boldsymbol{H}}^2$) if and only if ϕ is an inner function.

(ii) Show that $\{1, \phi, \phi^2, \phi^3, \dots\}$ is an orthonormal basis for $\widetilde{\boldsymbol{H}}^2$ if and only if there is a λ of modulus 1 such that $\phi(z) = \lambda z$.

2.5. Show that the linear span of $\{1\} \cup \{x^{r_n}\}$ is uniformly dense in $C[0,1]$ in each of the following cases:

(i) $r_n = 7n$ (this case can be done without using the Müntz–Szasz theorem).

(ii) $r_n = \sqrt{p_n + 1}$, where p_n is the nth prime number.

(iii) $r_n = \log^2(n+1)$.

2.6. Show that f is an outer function if both f and $1/f$ are in $\boldsymbol{H}^2$.

2.7. Prove that $1 + \phi$ is an outer function whenever ϕ is a function in $\boldsymbol{H}^\infty$ with $\|\phi\|_\infty < 1$.

2.8. Suppose f is in $\boldsymbol{H}^2$ and $\operatorname{Re} f(z) > 0$ for all $z \in \mathbb{D}$. Prove that f is an outer function.

2.9. Find the inner-outer factorization of $z - \lambda$ in each of the following cases:

(i) $|\lambda| < 1$.

(ii) $|\lambda| = 1$.

(iii) $|\lambda| > 1$.

2.10. Prove that polynomials do not have nonconstant singular inner factors.

2.11. Prove that every function f in $\boldsymbol{H}^2$ can be written in the form $f = f_1 + f_2$, where neither f_1 nor f_2 has a zero in $\mathbb{D}$ and $\|f_1\| \leq \|f\|$ and $\|f_2\| \leq \|f\|$.

2.12. Show that f in $\boldsymbol{H}^2$ implies that $|f(z)|^2$ has a harmonic majorant (i.e., that there exists a function u harmonic on $\mathbb{D}$ such that $|f(z)|^2 \leq u(z)$ for all $z \in \mathbb{D}$). Note that the converse of this is Exercise 1.6.

2.13. Let ϕ be a function in $\boldsymbol{L}^2$ such that $\log|\phi|$ is in $\boldsymbol{L}^1$. Define

$$F(z) = \exp\left(\frac{1}{2\pi}\int_0^{2\pi} \frac{e^{i\theta}+z}{e^{i\theta}-z} \log|\phi(e^{i\theta})|\, d\theta\right).$$

Show that F is an outer function such that $|\widetilde{F}(e^{i\theta}) = |\phi(e^{i\theta})|$. (Hint: See the proofs of Theorem 2.7.4, Corollary 2.7.5, and Theorem 2.7.7.)

2.14. Assume that ϕ and $1/\phi$ are both in $\boldsymbol{L}^\infty$. Show that there exists an outer function g in $\boldsymbol{H}^\infty$ such $|\widetilde{g}(e^{i\theta})| = |\phi(e^{i\theta})|$ a.e. Show also that $1/g$ is in $\boldsymbol{H}^\infty$.

2.15. *(i)* Show that the set of inner functions is a closed subset of $\boldsymbol{H}^2$.

(ii) Show that the set consisting of all singular inner functions and all constant inner functions is a closed subset of $\boldsymbol{H}^2$.

2.16. Let S be a singular inner function such that

$$\lim_{r\to 1^-} \frac{1}{2\pi}\int_0^{2\pi} \log|S(re^{i\theta})|\, d\theta = 0.$$

Prove that S is the constant function 1.

2.17. Let ϕ be an inner function satisfying

$$\lim_{r\to 1^-} \frac{1}{2\pi}\int_0^{2\pi} \log|\phi(re^{i\theta})|\, d\theta = 0.$$

Show that ϕ is a Blaschke product.

2.18. If ϕ is an inner function and α is in $\mathbb{D}$, show that the function

$$\frac{\alpha - \phi}{1 - \overline{\alpha}\phi}$$

is inner.

2.19. (Frostman's Theorem) Let ϕ be an inner function. Show that, for almost all α in $\mathbb{D}$ (with respect to area measure on $\mathbb{D}$), the function

$$\frac{\alpha - \phi}{1 - \overline{\alpha}\phi}$$

is a Blaschke product. (Hint: The three preceding exercises can be useful in establishing this result.)

2.20. Suppose that $\mathcal{M}_1$ and $\mathcal{M}_2$ are invariant subspaces of the unilateral shift U such that $\mathcal{M}_1 \subset \mathcal{M}_2$ and $\dim(\mathcal{M}_2 \ominus \mathcal{M}_1) > 1$. Show that there exists $\mathcal{M}$ invariant under U satisfying $\mathcal{M}_1 \subsetneq \mathcal{M} \subsetneq \mathcal{M}_2$.

2.9 Notes and Remarks

The early studies of functions in $\boldsymbol{H}^2$ were primarily concerned with geometric properties of individual analytic functions; the books of Privalov [39] and Goluzin [23] provide expositions from that point of view. The interest in $\boldsymbol{H}^2$ as a Hilbert space was largely stimulated by Beurling's 1949 paper [64] in which he characterized the invariant subspaces of the unilateral shift (Theorem 2.2.12). Beurling's paper stimulated functional analysts to study various properties of $\boldsymbol{H}^2$ as a linear space and to investigate operators on $\boldsymbol{H}^2$.

The F. and M. Riesz theorem (Theorem 2.3.3) was first established by the Riesz brothers in 1916 [145]; this paper also contains several deeper results. The definition of Blaschke products and the proof of their convergence is due to Blaschke [65]. The factorization into inner and outer factors (Theorem 2.3.4) was discovered by F. Riesz [144], although the terminology "inner" and "outer" is due to Beurling [64].

The Müntz–Szász theorem (Theorem 2.5.1) is due to Müntz [119] and Szász [161]. A number of variants and generalizations have been obtained; see [91] and the references it contains.

The proof we have presented of Euler's classical theorem [90] that the series of prime reciprocals diverges is a variant of proofs due to Bellman [63] and Clarkson [71] (also see [4, p. 18]). An entirely different proof can be found in [26, p. 17].

Herglotz's theorem (Theorem 2.6.3) was established in [106]. The representation theorem for singular inner functions (Theorem 2.6.5) is due to Smirnov [159], as is the representation of outer functions (Corollary 2.7.8) and the resulting complete factorization of functions in $\boldsymbol{H}^2$ (Corollary 2.7.9).

The study of attainable lattices (Definition 2.6.12) was initiated in [146]. Additional examples of attainable lattices can be found in [15] and [41].

Example 2.6.14 was pointed out by Sarason [148], who used it to give an alternative proof of the result of Dixmier [84] that the invariant subspace lattice of the "indefinite integral" operator (usually called the "Volterra operator") is isomorphic to the interval $[0,1]$. Exercise 2.4(ii) is from [32, p. 119, ex. 3], while Exercises 2.7 and 2.11 are from [35, pp. 27, 38].

Frostman's theorem is slightly sharper than what we have stated in Exercise 2.19; he proved [93] that, for ϕ an inner function, the set of α such that $\frac{\alpha-\phi}{1-\overline{\alpha}\phi}$ is not a Blaschke product has logarithmic capacity zero (for a proof of this see Garnett [22, p. 79]). A different proof of the almost everywhere version of Frostman's theorem is in [35, p. 45].

Chapter 3

Toeplitz Operators

The most-studied and best-known operators on the Hardy–Hilbert space are the Toeplitz operators. The forward and backward unilateral shifts are simple examples of Toeplitz operators; more generally, the Toeplitz operators are those operators whose matrices with respect to the standard basis of $\boldsymbol{H}^2$ have constant diagonals. We discuss many interesting results about spectra and other aspects of Toeplitz operators.

3.1 Toeplitz Matrices

Every bounded linear operator on a Hilbert space has a matrix representation with respect to each orthonormal basis of the space.

Definition 3.1.1. If A is a bounded linear operator on a Hilbert space $\mathcal{H}$ and $\{e_n\}_{n\in\mathcal{I}}$ is an orthonormal basis for $\mathcal{H}$, then *the matrix of A with respect to the given basis* is the matrix whose entry in position (m,n) for $m,n \in \mathcal{I}$ is (Ae_n, e_m).

It is easily seen that, just as in the familiar case of operators on finite-dimensional spaces, the effect of the operator A on $\sum_{n\in\mathcal{I}} c_n e_n$ can be obtained by multiplying the column vector $(c_n)_{n\in\mathcal{I}}$ on the left by the matrix of A (see, for example, [27, p. 23]).

We shall see that Toeplitz operators have matrices that are easily obtained from matrices of multiplication operators, so we begin with the study of the matrices of multiplication operators on $\boldsymbol{L}^2$ of the circle with respect to the standard basis $\{e^{in\theta}\}_{n=-\infty}^{\infty}$ for $\boldsymbol{L}^2$. Recall that M_ϕ denotes the operator on $\boldsymbol{L}^2$ consisting of multiplication by the $\boldsymbol{L}^\infty$ function ϕ (Definition 2.2.3).

Theorem 3.1.2. *Let ϕ be a function in $\boldsymbol{L}^\infty$ with Fourier series*

$$\sum_{n=-\infty}^{\infty} \phi_n e^{in\theta}.$$

Then the matrix of M_ϕ with respect to the orthonormal basis $\{e^{in\theta}\}_{n=-\infty}^{\infty}$ of $\boldsymbol{L}^2$ is

$$\begin{pmatrix} \ddots & \ddots & \ddots & & & & \\ \ddots & \phi_0 & \phi_{-1} & \phi_{-2} & & & \\ \ddots & \phi_1 & \phi_0 & \phi_{-1} & \phi_{-2} & & \\ & \phi_2 & \phi_1 & \boldsymbol{\phi_0} & \phi_{-1} & \phi_{-2} & \\ & & \phi_2 & \phi_1 & \phi_0 & \phi_{-1} & \ddots \\ & & & \phi_2 & \phi_1 & \phi_0 & \ddots \\ & & & & \ddots & \ddots & \ddots \end{pmatrix},$$

where boldface represents the $(0,0)$ *position.*

Proof. We compute, for each pair of integers (m,n),

$$(M_\phi e_n, e_m) = \frac{1}{2\pi}\int_0^{2\pi} \phi(e^{i\theta}) e^{in\theta}\overline{e^{im\theta}}\, d\theta = \frac{1}{2\pi}\int_0^{2\pi} \phi(e^{i\theta}) e^{-i(m-n)\theta}\, d\theta.$$

Therefore, for each integer k, the entry of the matrix for M_ϕ in position (m,n) is ϕ_k whenever $m-n=k$. This gives the result. □

Thus the matrices representing multiplication operators with respect to the standard basis for $\boldsymbol{L}^2$ are doubly infinite matrices whose diagonals are constant. Each such matrix is an example of a Toeplitz matrix.

Definition 3.1.3. A finite matrix, or a doubly infinite matrix (i.e., a matrix with entries in positions (m,n) for m and n integers), or a singly infinite matrix (i.e., a matrix with entries in positions (m,n) for m and n nonnegative integers) is called a *Toeplitz matrix* if its entries are constant along each diagonal. That is, the matrix $(a_{m,n})$ is Toeplitz if $a_{m_1,n_1} = a_{m_2,n_2}$ whenever $m_1 - n_1 = m_2 - n_2$.

We shall see that doubly infinite Toeplitz matrices are simpler than singly infinite ones. Theorem 3.1.2 shows that the matrices of multiplication operators on $\boldsymbol{L}^2$ are doubly infinite Toeplitz matrices; the converse also holds.

Theorem 3.1.4. *A bounded linear operator on $\boldsymbol{L}^2$ is multiplication by an $\boldsymbol{L}^\infty$ function if and only if its matrix with respect to the standard basis in $\boldsymbol{L}^2$ is a Toeplitz matrix.*

Proof. It was shown in Theorem 3.1.2 that every multiplication on $\boldsymbol{L}^2$ has a Toeplitz matrix. To establish the converse, assume that A has a Toeplitz matrix.

To show that $A = M_\phi$ for some $\phi \in \boldsymbol{L}^\infty$, it suffices, by Theorem 2.2.5, to show that $AW = WA$. This is equivalent to showing that, for all integers m and n, $(AWe_n, e_m) = (WAe_n, e_m)$. Observe that

$$(AWe_n, e_m) = (Ae_{n+1}, e_m) = (Ae_n, e_{m-1}),$$

by the assumption that the matrix of A has constant diagonals. Then

$$(Ae_n, e_{m-1}) = (Ae_n, W^* e_m) = (WAe_n, e_m).$$

Therefore $(AWe_n, e_m) = (WAe_n, e_m)$. □

The spectra and approximate point spectra of Toeplitz matrices on $\boldsymbol{L}^2$ are now easy to calculate. We need the following definition.

Definition 3.1.5. For $\phi \in \boldsymbol{L}^\infty$, the *essential range* of ϕ is defined to be

$$\operatorname{ess\,ran} \phi = \left\{\lambda : m\left\{e^{i\theta} : |\phi(e^{i\theta}) - \lambda| < \varepsilon\right\} > 0, \text{ for all } \varepsilon > 0\right\},$$

where m is the (normalized) Lebesgue measure.

Note that the essential norm of an $\boldsymbol{L}^\infty$ function ϕ (Definition 1.1.23) is equal to

$$\sup\{|\lambda| : \lambda \in \operatorname{ess\,ran} \phi\}.$$

Theorem 3.1.6. *If $\phi \in \boldsymbol{L}^\infty$, then $\sigma(M_\phi) = \Pi(M_\phi) = \operatorname{ess\,ran} \phi$.*

Proof. We prove this in two steps. We first show that $\operatorname{ess\,ran} \phi \subset \Pi(M_\phi)$, and then show that $\sigma(M_\phi) \subset \operatorname{ess\,ran} \phi$. These two assertions together imply the theorem.

Let $\lambda \in \operatorname{ess\,ran} \phi$. For each natural number n, define

$$E_n = \left\{e^{i\theta} : |\phi(e^{i\theta}) - \lambda| < \frac{1}{n}\right\}$$

and let χ_n be the characteristic function of E_n. Notice that $m(E_n) > 0$. Then

$$\begin{aligned}\|(M_\phi - \lambda)\chi_n\|^2 &= \frac{1}{2\pi}\int_0^{2\pi} |(\phi(e^{i\theta}) - \lambda)\chi_n(e^{i\theta})|^2 \, d\theta \\ &= \frac{1}{2\pi}\int_{E_n} |(\phi(e^{i\theta}) - \lambda)|^2 \, d\theta \\ &\leq \frac{1}{n^2} m(E_n).\end{aligned}$$

Also,

$$\|\chi_n\|^2 = \frac{1}{2\pi}\int_0^{2\pi} |\chi_n(e^{i\theta})|^2 \, d\theta = m(E_n) \neq 0.$$

Thus, if we define $f_n = \chi_n/\|\chi_n\|$, then $\{f_n\}$ is a sequence of unit vectors such that

$$\|(M_\phi - \lambda)f_n\| \leq \frac{1}{n}.$$

Therefore $\lambda \in \Pi(M_\phi)$.

Now suppose $\lambda \notin \operatorname{ess\,ran} \phi$. Then there exists $\varepsilon > 0$ such that

$$m\{e^{i\theta} \; : \; |\phi(e^{i\theta}) - \lambda| < \varepsilon\} = 0.$$

This means that the function $1/(\phi - \lambda)$ is defined almost everywhere, and, in fact, $1/|\phi - \lambda| \leq 1/\varepsilon$ a.e. Thus $1/(\phi - \lambda) \in \boldsymbol{L}^\infty$. But then the operator $M_{\frac{1}{\phi-\lambda}}$ is bounded and is clearly the inverse of $M_\phi - \lambda$. Thus $\lambda \notin \sigma(M_\phi)$. □

3.2 Basic Properties of Toeplitz Operators

The Toeplitz operators are the "compressions" of the multiplication operators to the subspace $\widetilde{\boldsymbol{H}}^2$, defined as follows.

Definition 3.2.1. For each ϕ in $\boldsymbol{L}^\infty$, the *Toeplitz operator with symbol* ϕ is the operator T_ϕ defined by

$$T_\phi f = P\phi f$$

for each f in $\widetilde{\boldsymbol{H}}^2$, where P is the orthogonal projection of $\boldsymbol{L}^2$ onto $\widetilde{\boldsymbol{H}}^2$.

Theorem 3.2.2. *The matrix of the Toeplitz operator with symbol ϕ with respect to the basis $\{e^{in\theta}\}_{n=0}^\infty$ of $\widetilde{\boldsymbol{H}}^2$ is*

$$T_\phi = \begin{pmatrix} \phi_0 & \phi_{-1} & \phi_{-2} & \phi_{-3} & \\ \phi_1 & \phi_0 & \phi_{-1} & \phi_{-2} & \ddots \\ \phi_2 & \phi_1 & \phi_0 & \phi_{-1} & \ddots \\ \phi_3 & \phi_2 & \phi_1 & \phi_0 & \ddots \\ & \ddots & \ddots & \ddots & \ddots \end{pmatrix},$$

where ϕ_k is the kth Fourier coefficient of ϕ.

Proof. This can easily be computed in the same way as the corresponding result for multiplication operators (Theorem 3.1.2). Alternatively, since P is the projection onto $\widetilde{\boldsymbol{H}}^2$ and T_ϕ is defined on $\widetilde{\boldsymbol{H}}^2$, the matrix of T_ϕ is the lower right corner of the matrix of M_ϕ. That is, the lower right corner of

$$M_\phi = \left(\begin{array}{ccc|ccc} \ddots & \ddots & \ddots & & & \\ \ddots & \phi_0 & \phi_{-1} & \phi_{-2} & & \\ \ddots & \phi_1 & \phi_0 & \phi_{-1} & \phi_{-2} & \\ \hline & \phi_2 & \phi_1 & \boldsymbol{\phi_0} & \phi_{-1} & \phi_{-2} \\ & & \phi_2 & \phi_1 & \phi_0 & \phi_{-1} & \ddots \\ & & & \phi_2 & \phi_1 & \phi_0 & \ddots \\ & & & & \ddots & \ddots & \ddots \end{array}\right),$$

so

$$T_\phi = \begin{pmatrix} \phi_0 & \phi_{-1} & \phi_{-2} & \phi_{-3} & \\ \phi_1 & \phi_0 & \phi_{-1} & \phi_{-2} & \ddots \\ \phi_2 & \phi_1 & \phi_0 & \phi_{-1} & \ddots \\ \phi_3 & \phi_2 & \phi_1 & \phi_0 & \ddots \\ & \ddots & \ddots & \ddots & \ddots \end{pmatrix}.$$

□

Thus Toeplitz operators have singly infinite Toeplitz matrices. We will show that every singly infinite Toeplitz matrix that represents a bounded operator is the matrix of a Toeplitz operator (Theorem 3.2.6).

The most tractable Toeplitz operators are the analytic ones.

Definition 3.2.3. The Toeplitz operator T_ϕ is an *analytic Toeplitz operator* if ϕ is in $\widetilde{\boldsymbol{H}}^\infty$.

Note that if ϕ is in $\widetilde{\boldsymbol{H}}^\infty$, then $T_\phi f = P\phi f = \phi f$ for all $f \in \widetilde{\boldsymbol{H}}^2$. It is easily seen that the standard matrix representations of analytic Toeplitz operators are lower triangular matrices.

Theorem 3.2.4. *If T_ϕ is an analytic Toeplitz operator, then the matrix of T_ϕ with respect to the basis $\{e^{in\theta}\}_{n=0}^\infty$ is*

$$T_\phi = \begin{pmatrix} \phi_0 & 0 & 0 & 0 & 0 & \\ \phi_1 & \phi_0 & 0 & 0 & 0 & \ddots \\ \phi_2 & \phi_1 & \phi_0 & 0 & 0 & \ddots \\ \phi_3 & \phi_2 & \phi_1 & \phi_0 & 0 & \ddots \\ & \ddots & \ddots & \ddots & \ddots & \ddots \end{pmatrix},$$

where $\phi(e^{i\theta}) = \sum_{k=0}^{\infty} \phi_k e^{in\theta}$.

Proof. The Fourier coefficients of ϕ with negative indices are 0 since ϕ is in $\widetilde{\boldsymbol{H}}^2$, so this follows immediately from Theorem 3.2.2. □

The following theorem is analogous to Theorem 2.2.5.

Theorem 3.2.5. *The commutant of the unilateral shift acting on* $\widetilde{\boldsymbol{H}}^2$ *is*

$$\left\{T_\phi : \phi \in \widetilde{\boldsymbol{H}}^\infty\right\}.$$

Proof. First, every analytic Toeplitz operator commutes with the shift. This follows from the fact that the shift is $M_{e^{i\theta}}$ and, for every $f \in \widetilde{\boldsymbol{H}}^2$,

$$T_\phi M_{e^{i\theta}} f = \phi e^{i\theta} f = e^{i\theta} \phi f = M_{e^{i\theta}} T_\phi f.$$

The proof of the converse is very similar to the corresponding proof of Theorem 2.2.5. Suppose that $AU = UA$. Let $\phi = Ae_0$. Then $\phi \in \widetilde{\boldsymbol{H}}^2$ and, since $AU = UA$, we have, for each positive integer n,

$$Ae_n = AU^n e_0 = U^n Ae_0 = U^n \phi = e^{in\theta} \phi.$$

Thus, by linearity, $Ap = \phi p$ for every polynomial $p \in \widetilde{\boldsymbol{H}}^2$. For an arbitrary $f \in \widetilde{\boldsymbol{H}}^2$, choose a sequence of polynomials $\{p_n\}$ such that $\{p_n\} \to f$ in $\widetilde{\boldsymbol{H}}^2$. Then, by continuity and the fact that $Ap_n = \phi p_n$, it follows that $\{\phi p_n\} \to Af$. Also, there exists a subsequence $\{p_{n_j}\}$ converging almost everywhere to f (since every sequence converging in $\boldsymbol{L}^2$ has a subsequence converging almost everywhere [47, p. 68]). Therefore, $\{\phi p_{n_j}\} \to \phi f$ a.e. Thus $Af = \phi f$ a.e.

It remains to be shown that ϕ is essentially bounded. If $A = 0$ the result is trivial, so we may assume that $\|A\| \neq 0$. Define the measurable function ψ by $\psi = \phi / \|A\|$. Note that ψ is in $\widetilde{\boldsymbol{H}}^2$. Then

$$\psi f = \frac{\phi f}{\|A\|} = \frac{Af}{\|A\|}$$

for every f in $\widetilde{\boldsymbol{H}}^2$. It follows that $\|\psi f\| \leq \|f\|$ for all f in $\widetilde{\boldsymbol{H}}^2$. Taking f to be the constant function 1 and a trivial induction yield $\|\psi^n\| \leq 1$ for every natural number n. Suppose that there is a positive ε such that the set E defined by

$$E = \left\{e^{i\theta} : |\psi(e^{i\theta})| \geq 1 + \varepsilon\right\}$$

has positive measure. Then

$$\begin{aligned}\|\psi^n\| &= \frac{1}{2\pi}\int_0^{2\pi} |\psi(e^{i\theta})|^n \, d\theta \\ &\geq \frac{1}{2\pi}\int_E |\psi(e^{i\theta})|^n \, d\theta \\ &\geq \frac{1}{2\pi}\int_E (1+\varepsilon)^n \, d\theta \\ &= (1+\varepsilon)^n m(E).\end{aligned}$$

Hence $\|\psi^n\| \leq 1$ for all n implies that $|\psi(e^{i\theta})| \leq 1$ a.e. Therefore ψ, and also ϕ, are in $\boldsymbol{L}^\infty$. □

By the correspondence between $\widetilde{\boldsymbol{H}}^2$ and $\boldsymbol{H}^2$, we can regard analytic Toeplitz operators as operating on $\boldsymbol{H}^2$. Each function $\phi \in \widetilde{\boldsymbol{H}}^\infty$ corresponds to a function ϕ analytic on $\mathbb{D}$. Then $\lim_{r\to 1^-} \phi(re^{i\theta}) = \phi(e^{i\theta})$ a.e. Regarding T_ϕ as an operator on $\boldsymbol{H}^2$, it follows that

$$(T_\phi f)(z) = \phi(z)f(z) \quad \text{for } f \in \boldsymbol{H}^2 \text{ and } z \in \mathbb{D}.$$

That is, the analytic Toeplitz operator T_ϕ is simply multiplication by the analytic function ϕ on $\boldsymbol{H}^2$.

We now show that all bounded Toeplitz matrices with respect to the standard basis of $\widetilde{\boldsymbol{H}}^2$ are the matrices of Toeplitz operators.

Theorem 3.2.6. *The Toeplitz operators on $\widetilde{\boldsymbol{H}}^2$ are the operators whose matrices with respect to the basis $\{e^{in\theta}\}_{n=0}^\infty$ of $\widetilde{\boldsymbol{H}}^2$ are Toeplitz matrices.*

Proof. The fact that every Toeplitz operator has a Toeplitz matrix has already been established (Theorem 3.2.2).

To prove the converse, let A be a bounded operator on $\widetilde{\boldsymbol{H}}^2$ whose matrix with respect to the standard basis is a Toeplitz matrix. Let P denote the orthogonal projection from $\boldsymbol{L}^2$ onto $\widetilde{\boldsymbol{H}}^2$. It is clear that AP is a bounded operator on $\boldsymbol{L}^2$.

For each natural number n, define the operator A_n on $\boldsymbol{L}^2$ by

$$A_n = W^{*n}APW^n.$$

We will show that $\{A_n\}$ converges (in a certain sense) to a multiplication operator, and that A is the Toeplitz operator whose symbol is the corresponding $\boldsymbol{L}^\infty$ function.

Clearly, $\|A_n\| \leq \|A\|$, since $\|W\| = \|W^*\| = \|P\| = 1$ (in fact, one can easily see that $\|A_n\| = \|A\|$).

For each pair (s,t) of integers, we have

$$(A_n e_s, e_t) = (W^{*n} A P W^n e_s, e_t) = (A P e_{s+n}, e_{t+n}).$$

But for n sufficiently large (in fact, for $n \geq -s$), this expression equals (Ae_{s+n}, e_{t+n}) which, by hypothesis, is constant with respect to n for $n \geq -s$. By linearity, it then follows that, for each pair (p,q) of trigonometric polynomials, $(A_n p, q)$ is constant for sufficiently large n. Define the bilinear form Ψ on trigonometric polynomials by

$$\Psi(p,q) = \lim_{n\to\infty} (A_n p, q).$$

Since $|\Psi(p,q)| \leq \|A\|\,\|p\|\,\|q\|$ for all p and q, $\Psi(p,q)$ is a bounded bilinear form on the subset of $\boldsymbol{L}^2$ consisting of trigonometric polynomials. Thus Ψ extends to a bounded bilinear form on $\boldsymbol{L}^2$ and there is a bounded linear operator A_0 on $\boldsymbol{L}^2$ such that $\Psi(f,g) = (A_0 f, g)$ for all f and g in $\boldsymbol{L}^2$. It follows that

$$\lim_{n\to\infty} (A_n f, g) = (A_0 f, g)$$

for all f and g in $\boldsymbol{L}^2$.

To finish the proof, it will be shown that A_0 is a multiplication operator and that the restriction of PA_0 to $\widetilde{\boldsymbol{H}}^2$ is A.

Observe that $(W^* A_n W f, g) = (A_{n+1} f, g)$, and also

$$(W^* A_n W f, g) = (A_n W f, W g).$$

Taking limits as n approaches infinity yields

$$(A_0 f, g) = (A_0 W f, W g) = (W^* A_0 W f, g).$$

Therefore $A_0 = W^* A_0 W$, or $W A_0 = A_0 W$. Since A_0 commutes with W, it follows from Theorem 2.2.5 that $A_0 = M_\phi$ for some $\phi \in \boldsymbol{L}^\infty$.

We show that A is T_ϕ. Note first that, for s and t nonnegative integers,

$$(PM_\phi e_s, e_t) = (PA_0 e_s, e_t) = (A_0 e_s, e_t) = \lim_{n\to\infty} (A_n e_s, e_t).$$

Since, as shown above, $(A_n e_s, e_t) = (Ae_{s+n}, e_{t+n})$ for n sufficiently large, and since $(Ae_{s+n}, e_{t+n}) = (Ae_s, e_t)$ because A is a Toeplitz matrix, it follows that

$$(PM_\phi e_s, e_t) = (Ae_s, e_t).$$

Hence A is the restriction of PM_ϕ to $\widetilde{\boldsymbol{H}}^2$, which is T_ϕ. □

The unilateral shift U is the Toeplitz operator $T_{e^{i\theta}}$. An easy consequence of the above theorem is a useful alternative characterization of Toeplitz operators.

Corollary 3.2.7. *The operator T is a Toeplitz operator if and only if $U^*TU = T$, where U is the unilateral shift.*

Proof. Note that, for nonnegative integers n and m,

$$(U^*TUe_n, e_m) = (TUe_n, Ue_m) = (Te_{n+1}, e_{m+1}).$$

Thus if T is a Toeplitz operator, $(U^*TUe_n, e_m) = (Te_n, e_m)$ (since T has a Toeplitz matrix by Theorem 3.2.6), so $U^*TU = T$.

Conversely, if $U^*TU = T$, then a trivial induction yields

$$(Te_{n+k}, e_{m+k}) = (Te_n, e_m)$$

for all natural numbers k, so T has a Toeplitz matrix and therefore is a Toeplitz operator by Theorem 3.2.6. □

The following is a trivial but important fact.

Theorem 3.2.8. *The mapping $\phi \mapsto T_\phi$ is an injective, bounded, linear, adjoint-preserving (i.e., $T_\phi^* = T_{\overline{\phi}}$) mapping from $\mathbf{L}^\infty$ onto the space of Toeplitz operators regarded as a subspace of the algebra of bounded linear operators on $\widetilde{\boldsymbol{H}}^2$.*

Proof. The map is obviously linear, and

$$\|T_\phi\| = \|PM_\phi\| \leq \|M_\phi\| = \|\phi\|_\infty,$$

so the mapping is bounded. If T_ϕ and T_ψ are equal, then comparing their matrices shows that ϕ and ψ have the same Fourier coefficients. Therefore the mapping is injective.

To show that the mapping is adjoint-preserving, simply compute as follows: for f and $g \in \widetilde{\boldsymbol{H}}^2$,

$$\begin{aligned}
(T_\phi^* f, g) &= (f, T_\phi g) = (f, PM_\phi g) = (f, \phi g) \\
&= \frac{1}{2\pi}\int_0^{2\pi} f(e^{i\theta})\overline{\phi(e^{i\theta})g(e^{i\theta})}\, d\theta \\
&= \frac{1}{2\pi}\int_0^{2\pi} \overline{\phi(e^{i\theta})} f(e^{i\theta})\overline{g(e^{i\theta})}\, d\theta \\
&= (\overline{\phi} f, g) = (PM_{\overline{\phi}} f, g) \\
&= (T_{\overline{\phi}} f, g).
\end{aligned}$$

□

It will be shown below (Corollary 3.3.2) that the above mapping is an isometry; i.e., $\|T_\phi\| = \|\phi\|_\infty$.

Although the sum of two Toeplitz operators is obviously a Toeplitz operator (as follows immediately from the definition of Toeplitz operators or, equally easily, from the characterization of Toeplitz matrices), the corresponding result rarely holds for products. One case in which it does hold is when ϕ is in $\widetilde{\boldsymbol{H}}^\infty$. In that case, T_ϕ is simply the restriction of M_ϕ to $\widetilde{\boldsymbol{H}}^2$, so $T_\psi T_\phi$ is obviously equal to $T_{\psi\phi}$ for all $\psi \in \boldsymbol{L}^\infty$.

The following lemma will be useful in several contexts.

Lemma 3.2.9. *If T_ψ and T_ϕ are Toeplitz operators and U is the unilateral shift, then*

$$U^*T_\psi T_\phi U - T_\psi T_\phi = P(e^{-i\theta}\psi) \otimes P(e^{-i\theta}\overline{\phi}) \qquad \textit{(see Notation 1.2.27)},$$

where P is the orthogonal projection of $\boldsymbol{L}^2$ onto $\widetilde{\boldsymbol{H}}^2$.

Proof. Note that $I = UU^* + e_0 \otimes e_0$, where $e_0 \otimes e_0$ is the orthogonal projection from $\widetilde{\boldsymbol{H}}^2$ onto the constants. Therefore

$$\begin{aligned} U^*T_\psi T_\phi U &= U^*T_\psi(UU^* + e_0 \otimes e_0)T_\phi U \\ &= U^*T_\psi UU^*T_\phi U + U^*T_\psi\,(e_0 \otimes e_0)\,T_\phi U \\ &= T_\psi T_\phi + U^*T_\psi e_0 \otimes e_0 T_\phi U \quad \text{(by Corollary 3.2.7)}. \end{aligned}$$

But $U^*T_\psi(e_0 \otimes e_0)T_\phi U = (U^*T_\psi e_0) \otimes (U^*T_{\overline{\phi}} e_0)$, by Theorem 1.2.28.

Notice that

$$U^*T_\psi e_0 = (T_{\overline{\psi}}U)^*e_0 = \left(T_{e^{i\theta}\overline{\psi}}\right)^* e_0 = T_{e^{-i\theta}\psi}e_0 = P(e^{-i\theta}\psi).$$

A similar computation applied to $U^*T_{\overline{\phi}}e_0$ yields the result. □

Definition 3.2.10. The Toeplitz operator T_ϕ is said to be *coanalytic* if T_ϕ^* is analytic (which is equivalent to saying that $\overline{\phi} \in \widetilde{\boldsymbol{H}}^2$).

Theorem 3.2.11. *For ψ and ϕ in $\boldsymbol{L}^\infty$, $T_\psi T_\phi$ is a Toeplitz operator if and only if either T_ψ is coanalytic or T_ϕ is analytic. In both of those cases, $T_\psi T_\phi = T_{\psi\phi}$.*

Proof. That $T_\psi T_\phi = T_{\psi\phi}$ when T_ϕ is analytic is trivial, as remarked above. If T_ψ is coanalytic, then

$$(T_\psi T_\phi)^* = T_\phi^* T_\psi^* = T_{\overline{\phi}} T_{\overline{\psi}} = T_{\overline{\phi}\,\overline{\psi}} \quad \text{(since } T_{\overline{\psi}} \text{ is analytic).}$$

Therefore

$$T_\psi T_\phi = \left(T_{\overline{\phi}\,\overline{\psi}}\right)^* = T_{\psi\phi}.$$

For the other direction, suppose that $T_\psi T_\phi$ is a Toeplitz operator. By Lemma 3.2.9,

$$U^* T_\psi T_\phi U - T_\psi T_\phi = P(e^{-i\theta}\psi) \otimes P(e^{-i\theta}\overline{\phi}).$$

The assumption that $T_\psi T_\phi$ is Toeplitz implies that

$$U^* T_\psi T_\phi U = T_\psi T_\phi$$

(by Corollary 3.2.7), and therefore

$$P(e^{-i\theta}\psi) \otimes P(e^{-i\theta}\overline{\phi}) = 0.$$

Thus at least one of $P(e^{-i\theta}\psi)$ or $P(e^{-i\theta}\overline{\phi})$ is 0. If $P(e^{-i\theta}\psi) = 0$, then $e^{-i\theta}\psi$ is orthogonal to $\widetilde{\boldsymbol{H}}^2$, so the Fourier coefficients of ψ corresponding to positive indices are all 0 and T_ψ is coanalytic. If $P(e^{-i\theta}\overline{\phi}) = 0$, then the Fourier coefficients of $\overline{\phi}$ corresponding to positive indices are all 0, from which it follows that T_ϕ is analytic. □

Corollary 3.2.12. *The product of two Toeplitz operators is* 0 *if and only if one of the factors is* 0.

Proof. Assume that $T_\psi T_\phi = 0$. Since 0 is a Toeplitz operator, the previous theorem implies that either T_ψ is coanalytic or T_ϕ is analytic, and that $T_\psi T_\phi = T_{\psi\phi} = 0$. Hence $\psi\phi = 0$.

If T_ϕ is analytic and not 0, then the measure of $\{e^{i\theta} : \phi(e^{i\theta}) = 0\}$ is 0 by the F. and M. Riesz theorem (Theorem 2.3.3), so $\psi = 0$ a.e. and therefore $T_\psi = 0$.

Similarly, if $T_{\overline{\psi}}$ is analytic and not 0, then the measure of $\{e^{i\theta} : \overline{\psi(e^{i\theta})} = 0\}$ is 0 by the F. and M. Riesz theorem (Theorem 2.3.3), so $\phi = 0$ a.e. and therefore $T_\phi = 0$. □

Similar techniques can be used to answer the question of when two Toeplitz operators commute. Notice that if both T_ϕ and T_ψ are analytic Toeplitz operators, then $T_\phi T_\psi = T_{\phi\psi} = T_{\psi\phi} = T_\psi T_\phi$, by the previous theorem. The same situation obtains if both T_ϕ and T_ψ are coanalytic. There is only one other case.

Theorem 3.2.13. *Let ϕ and ψ in $\mathbf{L}^\infty$. Then $T_\phi T_\psi = T_\psi T_\phi$ if and only if at least one of the following holds:*

(i) both T_ϕ and T_ψ are analytic;

(ii) both T_ϕ and T_ψ are coanalytic;

(iii) there exist complex numbers a and b, not both zero, such that $a\phi + b\psi$ is a constant.

Proof. It is clear that each of the three conditions implies commutativity.

For the converse, assume that $T_\phi T_\psi = T_\psi T_\phi$. By Lemma 3.2.9, we have

$$U^* T_\phi T_\psi U - T_\phi T_\psi = P(e^{-i\theta}\phi) \otimes P(e^{-i\theta}\overline{\psi})$$

and

$$U^* T_\psi T_\phi U - T_\psi T_\phi = P(e^{-i\theta}\psi) \otimes P(e^{-i\theta}\overline{\phi}).$$

Hence

$$P(e^{-i\theta}\phi) \otimes P(e^{-i\theta}\overline{\psi}) = P(e^{-i\theta}\psi) \otimes P(e^{-i\theta}\overline{\phi}).$$

There are two cases that we need to consider: either at least one of the vectors in the above equation is 0, or none of them are 0.

Assume first that one of the vectors is 0. We have four subcases. If $P(e^{-i\theta}\phi) = 0$, then, by Theorem 1.2.28, either $P(e^{-i\theta}\psi)$ or $P(e^{-i\theta}\overline{\phi})$ is 0. If $P(e^{-i\theta}\phi) = P(e^{-i\theta}\psi) = 0$, then the Fourier coefficients of ϕ and ψ corresponding to positive indices are all 0, and hence T_ϕ and T_ψ are both coanalytic Toeplitz operators and condition *(ii)* holds. If $P(e^{-i\theta}\phi) = P(e^{-i\theta}\overline{\phi}) = 0$, then ϕ must be a constant function and condition *(iii)* holds with $a = 1$ and $b = 0$.

If $P(e^{-i\theta}\overline{\psi}) = 0$, then, by Theorem 1.2.28, either $P(e^{-i\theta}\psi)$ or $P(e^{-i\theta}\overline{\phi})$ is 0. If $P(e^{-i\theta}\overline{\psi}) = P(e^{-i\theta}\psi) = 0$, then ψ must be a constant function and condition *(iii)* holds with $a = 0$ and $b = 1$. If $P(e^{-i\theta}\overline{\psi}) = P(e^{-i\theta}\overline{\phi}) = 0$, then the Fourier coefficients of ψ and ϕ corresponding to negative indices are all 0, and hence T_ψ and T_ϕ are both analytic Toeplitz operators and condition *(i)* holds. The other two subcases are very similar.

Assume that none of $P(e^{-i\theta}\phi)$, $P(e^{-i\theta}\overline{\psi})$, $P(e^{-i\theta}\psi)$, and $P(e^{-i\theta}\overline{\phi})$ is 0. Since

$$P(e^{-i\theta}\phi) \otimes P(e^{-i\theta}\overline{\psi}) = P(e^{-i\theta}\psi) \otimes P(e^{-i\theta}\overline{\phi}),$$

Theorem 1.2.28 implies that there exists a nonzero constant b such that

$$P(e^{-i\theta}\phi) = b\, P(e^{-i\theta}\psi) \quad \text{and} \quad P(e^{-i\theta}\overline{\phi}) = \overline{b}\, P(e^{-i\theta}\overline{\psi}).$$

This implies that $P(e^{-i\theta}(\phi - b\psi)) = 0$ and $P(e^{-i\theta}(\overline{\phi} - \overline{b}\,\overline{\psi})) = 0$, and hence that the function $\phi - b\psi$ is constant. Thus condition *(iii)* holds. □

Corollary 3.2.14. *If two Toeplitz operators commute with each other and neither is a linear combination of the identity and the other operator, then their product is a Toeplitz operator.*

Proof. It follows from the previous theorem that either the first operator is analytic or the second operator is coanalytic. Theorem 3.2.11 finishes the proof. □

The above theorem also yields a characterization of normal Toeplitz operators. First observe the following.

Theorem 3.2.15. *A Toeplitz operator is self-adjoint if and only if its symbol is real-valued almost everywhere.*

Proof. This follows immediately from the fact that $T_\phi = T_\phi^*$ if and only if $\phi = \overline{\phi}$. □

Corollary 3.2.16. *The Toeplitz operator T_ϕ is normal if and only if there exist complex numbers c and d and a real-valued function ψ in $\boldsymbol{L}^\infty$ such that $\phi = c\psi + d$ a.e. That is, the only normal Toeplitz operators are affine functions of self-adjoint Toeplitz operators.*

Proof. Since $T_\psi^* = T_\psi$, it is clear that every operator of the given form is normal.

The converse follows easily from Theorem 3.2.13. To prove it, suppose that T_ϕ commutes with $T_\phi^* = T_{\overline{\phi}}$. At least one of the three cases of Theorem 3.2.13 holds. If either of the first two cases holds, the function ϕ is a constant and the result is immediate. In the third case, $a\phi + b\overline{\phi} = k$ for some constant k. Taking conjugates and adding the two equations yields

$$2\mathrm{Re}\left((a + \overline{b})\phi\right) = k + \overline{k}.$$

Let the imaginary part of the function $(a + \overline{b})\phi$ be the real-valued function ψ. Then

$$(a + \overline{b})\phi = \frac{k + \overline{k}}{2} + i\psi.$$

If $a + \overline{b} \neq 0$, this clearly gives the result. On the other hand, if $a = -\overline{b}$, then $a\phi - \overline{a}\overline{\phi} = k$, so the imaginary part of ϕ is a constant function. It follows that ϕ has the given form with $\psi = \mathrm{Re}\,\phi$. □

The two general classes of linear operators on Hilbert space that are best understood are the class of normal operators and the class of compact operators. The previous theorem shows that the class of normal Toeplitz operators is quite restricted. There are substantially fewer compact Toeplitz operators.

Theorem 3.2.17. *The only compact Toeplitz operator is* 0.

Proof. Let T_ϕ be a compact Toeplitz operator. Recall that compact operators map weakly convergent sequences to norm-convergent sequences [12, p. 173]. Therefore we have

$$\{(T_\phi e_{s+n}, e_{t+n})\} \to 0 \qquad \text{as } n \to \infty,$$

since $\{e_{s+n}\}$ converges weakly to zero as $n \to \infty$. However, since T_ϕ is Toeplitz, $(T_\phi e_{s+n}, e_{t+n}) = \phi_{s-t}$, where ϕ_k is the kth Fourier coefficient of ϕ. Hence $\phi_{s-t} = 0$ for all nonnegative integers s and t; i.e., $\phi_k = 0$ for all integers k, which implies that $\phi = 0$ a.e. Thus $T_\phi = 0$. □

In fact, Toeplitz operators cannot even get close to compact operators (see Corollary 3.3.4 below).

3.3 Spectral Structure

The study of the spectra of Toeplitz operators turns out to be extremely interesting. Of course, the spectrum of T_ϕ, like every other property of T_ϕ, is determined by the symbol ϕ. There is, however, no known way of expressing the spectrum of T_ϕ in terms of the symbol for general ϕ. The special cases that are understood include some very beautiful theorems such as, for example, a complete description of the spectrum in the case that ϕ is continuous (Theorem 3.3.18).

There is a useful general result that the approximate point spectrum of T_ϕ always includes the essential range of ϕ.

Theorem 3.3.1 (The Spectral Inclusion Theorem). *For all ϕ in $\mathbf{L}^\infty$, the spectrum of M_ϕ is contained in the spectrum of T_ϕ. More precisely,*

$$\operatorname{ess\,ran} \phi = \Pi(M_\phi) = \sigma(M_\phi) \subset \Pi(T_\phi) \subset \sigma(T_\phi).$$

Proof. It has already been shown (Theorem 3.1.6) that $\operatorname{ess\,ran} \phi = \Pi(M_\phi) = \sigma(M_\phi)$.

Assume $\lambda \in \Pi(M_\phi)$. Then there exists a sequence $\{f_n\}$ of functions in $\boldsymbol{L}^2$ with $\|f_n\| = 1$ such that

$$\{\|(M_\phi - \lambda)f_n\|\} \to 0 \qquad \text{as } n \to \infty.$$

By removing a "tail" of small norm and renormalizing, for each n there is a g_n of norm 1 that has only a finite number of nonzero Fourier coefficients corresponding to negative indices and satisfies $\|f_n - g_n\| \leq \frac{1}{n}$. It follows immediately that

$$\{\|(M_\phi - \lambda)g_n\|\} \to 0 \qquad \text{as } n \to \infty.$$

Since the bilateral shift, W, shifts the Fourier coefficients of $\boldsymbol{L}^2$ functions to the right, for each n there exists a positive integer M_n such that $W^{M_n}g_n$ is in $\widetilde{\boldsymbol{H}}^2$. Since W is unitary and commutes with M_ϕ, we have

$$\|(M_\phi - \lambda)W^{M_n}g_n\| = \|W^{M_n}(M_\phi - \lambda)g_n\| = \|(M_\phi - \lambda)g_n\|.$$

Also, $\|W^{M_n}g_n\| = \|g_n\| = 1$, because W is unitary.

For each n, define $h_n = W^{M_n}g_n$. Then each h_n is in $\boldsymbol{H}^2$, $\|h_n\| = 1$, and

$$\{\|(M_\phi - \lambda)h_n\|\} \to 0.$$

But

$$\|(T_\phi - \lambda)h_n\| = \|P(M_\phi - \lambda)h_n\| \leq \|(M_\phi - \lambda)h_n\|.$$

Therefore $\{\|(T_\phi - \lambda)h_n\|\} \to 0$, so $\lambda \in \Pi(T_\phi)$. □

The following corollary is very useful.

Corollary 3.3.2. *For ϕ in $\boldsymbol{L}^\infty$, $\|\phi\|_\infty = \|M_\phi\| = \|T_\phi\| = r(T_\phi)$ (where $r(T_\phi)$ is the spectral radius; see Definition 1.2.2).*

Proof. We have already shown that $\|\phi\|_\infty = \|M_\phi\|$ (see Theorem 2.2.4).

It is an easy consequence of the spectral radius formula that the spectral radius of a normal operator is equal to its norm (see, for example, [41, p. 11] or [55, p. 351]). Thus $\|M_\phi\| = r(M_\phi)$. By the spectral inclusion theorem (Theorem 3.3.1), $r(M_\phi) \leq r(T_\phi)$. The spectral radius of every operator is at most its norm (Theorem 1.2.4), so $r(T_\phi) \leq \|T_\phi\|$. Thus

$$\|M_\phi\| = r(M_\phi) \leq r(T_\phi) \leq \|T_\phi\|.$$

But, since $T_\phi = PM_\phi \big|_{\widetilde{\boldsymbol{H}}^2}$, we have $\|T_\phi\| \leq \|P\| \, \|M_\phi\| = \|M_\phi\|$. Therefore the above inequalities must be equalities. □

Recall that an operator is said to be *quasinilpotent* if its spectrum is $\{0\}$.

Corollary 3.3.3. *The only quasinilpotent Toeplitz operator is the operator* 0.

Proof. If $r(T_\phi) = 0$, the previous corollary gives $\|\phi\|_\infty = 0$, which implies that $\phi = 0$ a.e.; i.e., $T_\phi = 0$. □

The operator 0 gets as close to a Toeplitz operator as any compact operator can.

Corollary 3.3.4. *If ϕ is in $\boldsymbol{L}^\infty$ and K is compact, then $\|T_\phi - K\| \geq \|T_\phi\|$.*

Proof. Since $\|T_{e^{-in\theta}}\| = 1$ for each natural number n,

$$\begin{aligned} \|T_\phi - K\| &\geq \|T_{e^{-in\theta}}(T_\phi - K)\| \\ &= \|T_{e^{-in\theta}\phi} - T_{e^{-in\theta}}K\| \qquad \text{(by Theorem 3.2.11)} \\ &\geq \|T_{e^{-in\theta}\phi}\| - \|T_{e^{-in\theta}}K\|. \end{aligned}$$

Note that $\|T_{e^{-in\theta}\phi}\| = \|e^{-in\theta}\phi\|_\infty = \|\phi\|_\infty = \|T_\phi\|$ (by the previous corollary) for every n. Also, $T_{e^{-in\theta}}$ is simply U^{*n}, where U is the unilateral shift. Therefore

$$\{T_{e^{-in\theta}} f\} \to 0 \qquad \text{for each } f \in \widetilde{\boldsymbol{H}}^2.$$

Since K is compact, it follows that

$$\{T_{e^{-in\theta}} K\} \to 0.$$

(The sequence obtained by multiplying a sequence of operators that goes pointwise to 0 and a compact operator goes to 0 in norm; see Exercise 1.19 in Chapter 1.) □

A Toeplitz operator may have another "best compact approximant" in addition to 0.

Example 3.3.5. *There is a rank-one operator K such that $\|S^* - K\| = \|S^*\|$, where S^* is the backward unilateral shift.*

Proof. Let $K = e_0 \otimes e_1$. Clearly, $\|S^* - K\| = 1 = \|S^*\|$. □

As we saw in Chapter 1 (Theorem 1.2.12), the numerical range of a normal operator is the convex hull of its spectrum. Since M_ϕ is a normal operator, and since, by the spectral inclusion theorem (Theorem 3.3.1), we have $\sigma(M_\phi) \subset \sigma(T_\phi)$, the following theorem provides a more precise location of the spectrum.

Theorem 3.3.6. *For ϕ in $\boldsymbol{L}^\infty$, the following sets are identical:*

(i) the closed convex hull of $\sigma(T_\phi)$;

(ii) the closed convex hull of $\sigma(M_\phi)$;

(iii) the closure of the numerical range of T_ϕ;

(iv) the closure of the numerical range of M_ϕ;

(v) the closed convex hull of the essential range of ϕ.

Proof. The closure of the numerical range of M_ϕ is the convex hull of its spectrum, since M_ϕ is normal (Theorem 1.2.12). By the spectral inclusion theorem (Theorem 3.3.1), the closed convex hull of $\sigma(M_\phi)$ is contained in the closed convex hull of $\sigma(T_\phi)$, and the latter is contained in the closure of the numerical range of T_ϕ (Theorem 1.2.11). It is apparent that the numerical range of T_ϕ is contained in the numerical range of M_ϕ, since $(T_\phi f, f)$ for $f \in \widetilde{\boldsymbol{H}}^2$ is equal to $(M_\phi f, f)$.

It follows that the containments established above are actually equalities, which proves parts *(i)–(iv)* of the theorem. Part *(v)* follows from Theorem 3.1.6. □

The following easy corollary is quite useful.

Corollary 3.3.7. *For every ϕ in $\boldsymbol{L}^\infty$,* ess ran ϕ *is contained in $\sigma(T_\phi)$ and $\sigma(T_\phi)$ is contained in the closed convex hull of* ess ran ϕ. *In particular, if* ess ran ϕ *is convex, then $\sigma(T_\phi) =$* ess ran ϕ.

Proof. This is an immediate consequence of the previous theorem together with the spectral inclusion theorem (Theorem 3.3.1). □

The spectra of Toeplitz operators can also be computed in some cases in which ess ran ϕ is not convex. In particular, $\sigma(T_\phi)$ is easily described when T_ϕ is an analytic Toeplitz operator.

Theorem 3.3.8. *If $\phi \in \boldsymbol{H}^\infty$, then $\sigma(T_\phi)$ is the closure of $\phi(\mathbb{D})$.*

Proof. For the proof of this theorem, it is convenient to regard T_ϕ as acting on $\boldsymbol{H}^2$ rather than on $\widetilde{\boldsymbol{H}}^2$. To establish one inclusion, suppose that $\lambda = \phi(z_0)$ for some $z_0 \in \mathbb{D}$. Then

$$((T_\phi - \lambda)f)(z_0) = (\phi(z_0) - \lambda)f(z_0) = 0$$

for every $f \in \boldsymbol{H}^2$. This implies that the operator $T_\phi - \lambda$ is not surjective: all functions in its range have a zero at z_0. Therefore $T_\phi - \lambda$ cannot be invertible. This shows that

$$\phi(\mathbb{D}) \subset \sigma(T_\phi),$$

and thus that

$$\overline{\phi(\mathbb{D})} \subset \sigma(T_\phi).$$

Conversely, assume λ is not in the closure of $\phi(\mathbb{D})$. Define $\delta = \operatorname{dist}\left(\lambda, \overline{\phi(\mathbb{D})}\right)$; then $\delta > 0$. Since $|\phi(z) - \lambda| \geq \delta$ for all $z \in \mathbb{D}$, $1/(\phi(z) - \lambda)$ is analytic and bounded by $1/\delta$ on $\mathbb{D}$. It is therefore obvious that

$$(T_\phi - \lambda)^{-1} = T_{\frac{1}{\phi - \lambda}},$$

so $\lambda \notin \sigma(T_\phi)$. □

A corresponding result holds for coanalytic Toeplitz operators.

Corollary 3.3.9. *If T_ϕ is a coanalytic Toeplitz operator, and if $\overline{\phi}$ is the function in $\boldsymbol{H}^\infty$ whose boundary function is the complex conjugate of ϕ a.e., then $\sigma(T_\phi)$ is the closure of the set of complex conjugates of $\overline{\phi}(\mathbb{D})$.*

Proof. Recall that the adjoint of a Toeplitz operator is the Toeplitz operator whose symbol is the complex conjugate of the given one (Theorem 3.2.8). Since the complex conjugate of ϕ is in $\widetilde{\boldsymbol{H}}^2$, the result follows from the previous theorem and the fact that the spectrum of the adjoint of an operator consists of the complex conjugates of the elements of the spectrum of the operator (Theorem 1.2.4). □

The following easily proven theorem has several useful consequences.

Theorem 3.3.10 (The Coburn Alternative). *If ϕ is a function in $\boldsymbol{L}^\infty$ other than 0, then at least one of T_ϕ and T_ϕ^* is injective.*

Proof. Suppose that $T_\phi f = 0$ for some $f \neq 0$. Suppose also that

$$T_\phi^* g = P(\overline{\phi} g) = 0.$$

It must be shown that $g = 0$.

We are given $P(\phi f) = 0$, where P is the projection of $\boldsymbol{L}^2$ onto $\widetilde{\boldsymbol{H}}^2$. Then the functions ϕf and $\overline{\phi} g$ are both in $(\widetilde{\boldsymbol{H}}^2)^\perp$. It follows that the only Fourier coefficients of $\overline{\phi}\,\overline{f}$ and $\phi \overline{g}$ that might be different from 0 are those in positions $\{1, 2, 3, \ldots\}$. Since f and g are in $\widetilde{\boldsymbol{H}}^2$, this implies that the functions $\overline{\phi}\,\overline{f} g$

and $\phi\overline{g}f$ in $\boldsymbol{L}^1(S^1)$ (recall that the product of two functions in an $\boldsymbol{L}^2$ space is in $\boldsymbol{L}^1$; see [47, p. 66]) have Fourier coefficients different from 0 at most in positions $\{1,2,3,\ldots\}$. But

$$\overline{\phi\overline{g}f} = \overline{\phi}\,\overline{f}g,$$

so $\phi\overline{g}f$ and its conjugate each have nonzero Fourier coefficients at most in positions $\{1,2,3,\ldots\}$. Therefore $\phi\overline{g}f$ is the constant function 0.

Since $f \neq 0$, the F. and M. Riesz theorem (Theorem 2.3.3) implies that the set on which f vanishes has measure 0. Therefore $\phi\overline{g}$ equals 0 almost everywhere. Since ϕ is not 0 there is a set of positive measure on which $\overline{g}$ vanishes. Hence g vanishes on a set of positive measure, and it is therefore 0 by the F. and M. Riesz theorem (Theorem 2.3.3). □

In the following sense, Toeplitz operators other than 0 always have some properties of invertibility.

Corollary 3.3.11. *A Toeplitz operator, other than* 0, *has dense range if it is not injective.*

Proof. If T_ϕ is not injective, it follows from the Coburn alternative (Theorem 3.3.10) that T_ϕ^* is injective. If the range of T_ϕ were not dense, there would exist a g different from 0 such that $(T_\phi f, g) = 0$ for all $f \in \widetilde{\boldsymbol{H}}^2$. But $(T_\phi f, g) = (f, T_\phi^* g)$, so, in particular, $(T_\phi^* g, T_\phi^* g) = 0$ and $T_\phi^* g = 0$. This would contradict the injectivity of T_ϕ^*. □

Corollary 3.3.12. *For* ϕ *a nonconstant function in* $\boldsymbol{L}^\infty$,

$$\Pi_0(T_\phi) \bigcap \overline{\Pi_0(T_\phi^*)} = \varnothing,$$

where $\overline{\Pi_0(T_\phi^*)}$ *denotes the set of complex conjugates of the eigenvalues of* T_ϕ^*.

Proof. Suppose that $\lambda \in \Pi_0(T_\phi)$. Then there is a function f other than zero such that $(T_\phi - \lambda)f = (T_{\phi-\lambda})\,f = 0$. Suppose that $(T_\phi^* - \overline{\lambda})g = 0$; it must be shown that g equals 0. Clearly $T_\phi^* - \overline{\lambda} = T_{\phi-\lambda}^*$. Thus the Coburn alternative (Theorem 3.3.10) implies that g is 0. □

Corollary 3.3.13. *If* ϕ *is a real-valued nonconstant function in* $\boldsymbol{L}^\infty$, *then* $\Pi_0(T_\phi) = \varnothing$.

Proof. In this case, T_ϕ is self-adjoint, so its spectrum is real. Thus if there was a λ in $\Pi_0(T_\phi)$, $\lambda = \overline{\lambda}$ would be in $\Pi_0(T_\phi^*)$, contradicting the previous corollary. □

We next determine the spectra of self-adjoint Toeplitz operators. Recall that T_ϕ is self-adjoint if and only if ϕ is real-valued a.e. (Theorem 3.2.15). We require a definition.

Definition 3.3.14. For ϕ a real-valued function in $\boldsymbol{L}^\infty$, the *essential infimum* of ϕ, denoted by ess inf ϕ, is the greatest lower bound of the essential range of ϕ, and the *essential supremum* of ϕ, denoted by ess sup ϕ, is the least upper bound of the essential range of ϕ.

Theorem 3.3.15. *If T_ϕ is self-adjoint, then $\sigma(T_\phi)$ is the interval*

$$\{t\,:\, \operatorname{ess\,inf}\,\phi \leq t \leq \operatorname{ess\,sup}\,\phi\}.$$

Proof. It is clear that the above interval is the closed convex hull of the essential range of ϕ, so it contains $\sigma(T_\phi)$ by Theorem 1.2.11.

For the opposite inclusion, first notice that, since ess ran $\phi \subset \sigma(T_\phi)$ by the spectral inclusion theorem (Theorem 3.3.1), and since ess ran ϕ contains ess inf ϕ and ess sup ϕ (because ess ran ϕ is closed), both ess inf ϕ and ess sup ϕ are in $\sigma(T_\phi)$.

Take λ in the open interval (ess inf ϕ, ess sup ϕ); we will show that $T_\phi - \lambda$ is not invertible. In fact, we will show that $T_\phi - \lambda$ is not surjective by showing that there is no $f \in \widetilde{\boldsymbol{H}}^2$ satisfying $(T_\phi - \lambda)f = 1$.

Suppose there was such an f. Then

$$P((\phi - \lambda)f) = 1,$$

or, equivalently,

$$P((\phi - \lambda)f - 1) = 0.$$

That is, $(\phi - \lambda)f - 1$ is in $\boldsymbol{L}^2 \ominus \widetilde{\boldsymbol{H}}^2$. But then

$$\overline{(\phi - \lambda)f - 1} \in \widetilde{\boldsymbol{H}}^2.$$

Since $\phi = \overline{\phi}$ and $\lambda \in \mathbb{R}$, we have

$$(\phi - \lambda)\overline{f} - 1 \in \widetilde{\boldsymbol{H}}^2,$$

and thus

$$(\phi - \lambda)\overline{f} \in \widetilde{\boldsymbol{H}}^2.$$

Since $f \in \widetilde{\boldsymbol{H}}^2$ implies that $\overline{f}e_n$ is in $\boldsymbol{L}^2 \ominus \boldsymbol{H}^2$ for $n = -1, -2, \ldots$, we then have

$$0 = ((\phi - \lambda)\overline{f}, \overline{f}e_n) = \frac{1}{2\pi} \int_0^{2\pi} \left((\phi(e^{i\theta}) - \lambda) \overline{f(e^{i\theta})} \right) \overline{\left(\overline{f(e^{i\theta})} e^{in\theta} \right)} \, d\theta.$$

This can be simplified to yield

$$0 = \frac{1}{2\pi} \int_0^{2\pi} (\phi(e^{i\theta}) - \lambda) \left| f(e^{i\theta}) \right|^2 e^{-in\theta} \, d\theta \quad \text{for } n = -1, -2, \ldots .$$

Taking complex conjugates, we get

$$0 = \frac{1}{2\pi} \int_0^{2\pi} (\phi(e^{i\theta}) - \lambda) \left| f(e^{i\theta}) \right|^2 e^{in\theta} \, d\theta \quad \text{for } n = -1, -2, \ldots .$$

Thus

$$0 = \frac{1}{2\pi} \int_0^{2\pi} (\phi(e^{i\theta}) - \lambda) \left| f(e^{i\theta}) \right|^2 e^{in\theta} \, d\theta \quad \text{for } n = \pm 1, \pm 2, \ldots .$$

It follows that the function $(\phi - \lambda)|f|^2$ must be a (real) constant c, since all its Fourier coefficients, except possibly the zeroth, are 0. Note that $(\phi - \lambda)f \neq 0$, since $P((\phi - \lambda)f) = 1$. Therefore c cannot be 0 ($\overline{f} \neq 0$ a.e. by the F. and M. Riesz theorem, Theorem 2.3.3).

Thus $(\phi - \lambda)|f|^2$ is a nonzero constant. But this is impossible. Indeed, since $\lambda \in (\operatorname{ess\,inf} \phi, \operatorname{ess\,sup} \phi)$, we have $\phi(e^{i\theta}) - \lambda > 0$ for $e^{i\theta}$ in a set of positive measure and also $\phi(e^{i\theta}) - \lambda < 0$ on a set of positive measure. Therefore $(\phi - \lambda)|f|^2$ takes both positive and negative values on (different) sets of positive measure. But that cannot occur, since $(\phi - \lambda)|f|^2$ is a nonzero constant. This contradiction establishes the theorem. □

We next proceed to establish an elegant description of the spectrum of the Toeplitz operator T_ϕ in the case that ϕ is a continuous function on S^1. The description requires the concept of the index (or winding number) of a function.

Definition 3.3.16. Let γ be a continuous complex-valued function on S^1 (i.e., γ is a closed curve), and let a be a point that is not in the range of γ. The *index* of the point a with respect to γ (also called the *winding number* of γ about a) is defined as

$$\operatorname{Ind}_a \gamma = \frac{1}{2\pi i} \int_\gamma \frac{1}{z - a} \, dz.$$

Recall the standard result from complex analysis that the index measures the number of times that the curve *winds* around the point a, positively in the counterclockwise direction and negatively in the clockwise direction ([9, p. 81], [47, p. 204]).

We need the following fundamental lemma concerning the relationship between the index of the product of two curves and the indices of the factors.

Lemma 3.3.17. *Let β and γ be continuous functions on S^1 neither of which assumes the value* 0. *Then*

$$\mathrm{Ind}_0\,(\beta\gamma) = \mathrm{Ind}_0\,\beta + \mathrm{Ind}_0\,\gamma.$$

Proof. See [9, p. 81]. □

The spectrum of a Toeplitz operator with continuous symbol is characterized as follows.

Theorem 3.3.18. *Let ϕ be a continuous function on S^1. Then*

$$\sigma(T_\phi) = \mathrm{ran}\ \phi \cup \{a \in \mathbb{C} : a \notin \mathrm{ran}\ \phi \ \ \textit{and}\ \ \mathrm{Ind}_a\,\phi \neq 0\},$$

where ran ϕ *is the range of* ϕ.

Proof. Note that ess ran ϕ = ran ϕ for continuous functions ϕ. Thus it follows from the spectral inclusion theorem (Theorem 3.3.1) that ran $\phi \subset \sigma(T_\phi)$.

Thus we only need to show that, for a not in ran ϕ, $T_\phi - a$ is invertible if and only if $\mathrm{Ind}_a\,\phi = 0$.

Since $a \notin \mathrm{ran}\ \phi$ if and only if $0 \notin \mathrm{ran}\ (\phi - a)$, $T_\phi - a = T_{\phi-a}$, and $\mathrm{Ind}_a\,\phi = \mathrm{Ind}_0\,(\phi - a)$, we may assume that $a = 0$. We make that assumption; i.e., we assume that $\phi(e^{i\theta}) \neq 0$ for all θ. In this case we must show that T_ϕ is invertible if and only if $\mathrm{Ind}_0\,\phi = 0$.

Since ϕ is continuous, there exists a $\delta > 0$ such that $|\phi(e^{i\theta})| > \delta$ for all θ. Choose a trigonometric polynomial p such that $|p(e^{i\theta}) - \phi(e^{i\theta})| < \delta/3$ for all θ (it is a very well-known elementary fact that the trigonometric polynomials are uniformly dense in the space of continuous functions of the circle; see [47, p. 91]). It is easy to see that $|p(e^{i\theta})| > 2\delta/3$ for all θ. Therefore we may define a continuous function ψ by $\psi = (\phi - p)/p$, so that

$$\phi = p(1 + \psi).$$

It is obvious that $|\psi(e^{i\theta})| \leq \frac{1}{2}$. Thus ran $(1 + \psi)$ is contained in the disk $\{z : |z - 1| \leq \frac{1}{2}\}$. Hence the curve parametrized by $1 + \psi$ cannot wind around 0, so $\mathrm{Ind}_0\,(1 + \psi) = 0$. By the above lemma,

$$\mathrm{Ind}_0\,\phi = \mathrm{Ind}_0\,p + \mathrm{Ind}_0\,(1+\psi) = \mathrm{Ind}_0\,p.$$

It therefore suffices to show that T_ϕ is invertible if and only if $\mathrm{Ind}_0\,p = 0$.

Choose n sufficiently large that $e^{in\theta}p(e^{i\theta})$ has no nonzero Fourier coefficients with negative indices; that is, such that $e^{in\theta}p(e^{i\theta})$ is a polynomial in $e^{i\theta}$. Let q be defined by $q(e^{i\theta}) = e^{in\theta}p(e^{i\theta})$. We consider the polynomial $q(z)$ as a function on the complex plane. Notice that $q(z)$ has no zeros on S^1, since $|p(e^{i\theta})| > 2\delta/3$ for all θ. Let the zeros of $q(z)$ other than 0 inside the circle S^1 (repeated according to multiplicity) be the numbers $z_1, z_2, \ldots, z_k$, and let $w_1, w_2, \ldots, w_l$ be the zeros of $q(z)$ (repeated according to multiplicity) outside the circle S^1. Then,

$$q(z) = cz^m \prod_{j=1}^{k}(z-z_j)\prod_{j=1}^{l}(z-w_j),$$

where c is a constant other than 0 and m is a nonnegative integer.

On S^1, $z^{-1} = \overline{z}$. Thus we can rewrite the polynomial as a function on S^1 in the form

$$q(e^{i\theta}) = e^{in\theta}p(e^{i\theta}) = ce^{im\theta}e^{ik\theta}\prod_{j=1}^{k}(1-z_je^{-i\theta})\prod_{j=1}^{l}(e^{i\theta}-w_j).$$

Dividing by $e^{in\theta}$, we get

$$p(e^{i\theta}) = ce^{i(m-n+k)\theta}\prod_{j=1}^{k}(1-z_je^{-i\theta})\prod_{j=1}^{l}(e^{i\theta}-w_j).$$

Thus

$$\mathrm{Ind}_0\,p = \mathrm{Ind}_0\,(e^{i(m-n+k)\theta}) + \sum_{j=1}^{k}\mathrm{Ind}_0\,(1-z_je^{-i\theta}) + \sum_{j=1}^{l}\mathrm{Ind}_0\,(e^{i\theta}-w_j),$$

by repeated application of Lemma 3.3.17. Note that $\mathrm{Ind}_0\,(e^{i\theta}-w_j) = 0$ for each j, since w_j is outside S^1. Also,

$$\mathrm{Ind}_0\,(1-z_je^{-i\theta}) = \mathrm{Ind}_0\,(e^{-i\theta}(e^{i\theta}-z_j)) = \mathrm{Ind}_0\,e^{-i\theta} + \mathrm{Ind}_0\,(e^{i\theta}-z_j) = -1+1 = 0,$$

since $\mathrm{Ind}_0\,(e^{i\theta}-z_j) = 1$ because z_j is inside S^1.

Therefore, $\mathrm{Ind}_0\,p = \mathrm{Ind}_0\,(e^{i(m-n+k)\theta}) = m-n+k$. Thus it suffices to show that T_ϕ is invertible if and only if $m-n+k=0$.

We first consider the case $m-n+k \geq 0$. Since $\phi = p(1+\psi)$, we can write

$$\phi(e^{i\theta}) = c\left(\prod_{j=1}^{k}(1 - z_j e^{-i\theta})\right)(1 + \psi(e^{i\theta}))\left(\prod_{j=1}^{l}(e^{i\theta} - w_j)\right)e^{i(m-n+k)\theta}.$$

Since each of $T_{1-z_je^{-i\theta}}$ is coanalytic, each of $T_{e^{i\theta}-w_j}$ is analytic, and $T_{e^{i(m-n+k)\theta}}$ is analytic, we can factor T_ϕ as

$$T_\phi = c\left(\prod_{j=1}^{k} T_{1-z_je^{-i\theta}}\right) T_{1+\psi}\left(\prod_{j=1}^{l} T_{e^{i\theta}-w_j}\right) T_{e^{i(m-n+k)\theta}},$$

by Theorem 3.2.11.

Note that $T_{1+\psi}$ is invertible (by Theorem 1.2.4), since

$$\|1 - T_{1+\psi}\| = \|T_\psi\| \leq \|\psi\|_\infty \leq \frac{1}{2} < 1.$$

Also, each $T_{e^{i\theta}-w_j}$ is invertible by Theorem 3.3.8, and each $T_{1-z_je^{-i\theta}}$ is invertible by Corollary 3.3.9. Thus, in the case $m - n + k \geq 0$, T_ϕ is invertible if and only if $T_{e^{i(m-n+k)\theta}}$ is invertible. But it follows from Theorem 3.3.8 that $T_{e^{i(m-n+k)\theta}}$ is invertible if and only if $m - n + k = 0$.

We are left with the case $m - n + k < 0$; we must show that T_ϕ is not invertible in this case. The factorization of T_ϕ can be written

$$T_\phi = cT_{e^{-i(m-n+k)\theta}}\left(\prod_{j=1}^{k} T_{1-z_je^{-i\theta}}\right) T_{1+\psi}\left(\prod_{j=1}^{l} T_{e^{i\theta}-w_j}\right),$$

since $T_{e^{-i(m-n+k)\theta}}$ is coanalytic in this case. As in the previous case, it follows that T_ϕ is invertible if and only if $T_{e^{-i(m-n+k)\theta}}$ is invertible. However, Corollary 3.3.9 implies that $T_{e^{-i(m-n+k)\theta}}$ is not invertible. This completes the proof. □

3.4 Exercises

3.1. Show that $UU^* = I - e_0 \otimes e_0$.

3.2. Prove that $T_{1-w\overline{z}}$ is invertible when $|w| < 1$.

3.3. Let the operators $A_n = W^{*n}APW^n$ on $\boldsymbol{L}^2$ be defined as in the proof of Theorem 3.2.6. Prove that $\|A_n\| = \|A\|$ for every n.

3.4. Show that M_ϕ and T_ϕ have no eigenvalues in common if ϕ is a function other than zero in $\boldsymbol{L}^\infty$.

3.5. Let T_ϕ be an analytic Toeplitz operator. Show that every reproducing kernel function (Definition 1.1.7) is an eigenvector for T_ϕ^*.

3.6. Let ϕ be a real-valued function in $\boldsymbol{L}^\infty$. Prove that if the constant function 1 is in the range of T_ϕ, then 0 is not an interior point of the spectrum of T_ϕ. (Hint: See the proof of Theorem 3.3.15.)

3.7. Show that a Toeplitz operator is an isometry if and only if its symbol is an inner function. (See Exercise 2.4(i) in Chapter 2.)

3.8. Prove that the only unitary Toeplitz operators are the unimodular multiples of the identity.

3.9. Write out the details of "the other two subcases" in the proof of Theorem 3.2.13.

3.10. Prove that, for each ϕ in $\boldsymbol{L}^\infty$, $T_\phi U - UT_\phi$ has rank at most one.

3.11. Prove that $T_{\phi\psi} - T_\phi T_\psi$ is a compact operator if at least one of ϕ and ψ is the sum of a function in $\widetilde{\boldsymbol{H}}^\infty$ and a function continuous on S^1. (Hint: In the continuous case, this can be established by approximating the continuous function by trigonometric polynomials and using the previous exercise. It can then be shown that adding a function in $\widetilde{\boldsymbol{H}}^2$ does not change $T_{\phi\psi} - T_\phi T_\psi$).

3.12. Prove that if a nonzero Toeplitz operator T is not injective, then there is a function g in the range of T such that $g(0) \neq 0$. (Hint: Use Exercise 3.10.)

3.13. Show that if a nonzero Toeplitz operator is not injective, then the constant function 1 is in its range. (Hint: Use the previous exercise.)

3.14. Prove that if a nonzero Toeplitz operator is not injective, then its range includes all polynomials. (Hint: Use the previous exercise.)

3.15. Let A be a bounded operator and ϕ be in $\boldsymbol{H}^\infty$. Prove that A commutes with T_ϕ if and only if, for every λ in $\mathbb{D}$, the function $A^* k_\lambda$ is orthogonal to $(\phi - \phi(\lambda))\boldsymbol{H}^2$. (Recall that k_λ is the reproducing kernel for λ.)

3.16. Prove that a nonzero analytic Toeplitz operator does not have finite-dimensional invariant subspaces.

3.17. Prove that if K is compact and commutes with an analytic Toeplitz operator, then K is quasinilpotent.

3.18. Assume that ϕ and $1/\phi$ are in $\boldsymbol{L}^\infty$. Prove that T_ϕ is invertible if and only if $T_{\phi/|\phi|}$ is invertible. (Hint: Using Exercise 2.14, construct an invertible outer function g such that $|\widetilde{g}(e^{i\theta})| = |\phi(e^{i\theta})|^{1/2}$ a.e. Show that the result follows from the equation $T_\phi = T_f^* T_{\phi/|\phi|} T_f$.)

3.5 Notes and Remarks

The study of Toeplitz operators originates with Otto Toeplitz [162]. The exposition that we have presented of the relationship between the matrices and symbols of Laurent and Toeplitz matrices is due to Brown and Halmos [67]. The paper [67] was stimulated by the work of Hartman and Wintner [100]. Brown and Halmos [67] also contains Corollary 3.2.7, Theorem 3.2.8, Theorem 3.2.11, Corollary 3.2.12, Theorem 3.2.13, and Theorem 3.3.6.

The spectral inclusion theorem (Theorem 3.3.1) was discovered by Hartman and Wintner [100]. Theorem 3.3.10, the Coburn alternative, appears in [75]. Hartman and Wintner [101] were the first to determine the spectra of self-adjoint Toeplitz operators (Theorem 3.3.15). The beautiful theorem characterizing the spectra of Toeplitz operators with continuous symbols (Theorem 3.3.18) is the combined work of several mathematicians: Kreĭn [109], Calderón, Spitzer, and Widom [68], Widom [169], and Devinatz [83]. An extension of this result to functions that are the sum of a continuous function and a function in $\widetilde{\boldsymbol{H}}^\infty$ can be found in [16, Theorem 7.36]. The spectra of Toeplitz operators with piecewise continuous symbols are investigated in Widom [169], Devinatz [83], and Gohberg and Krupnik [94].

Widom [170], answering a question of Paul Halmos, proved that the spectrum of every Toeplitz operator is connected. A somewhat simplified exposition of the proof is given by Douglas [16]. (Douglas actually proves the stronger result that the essential spectra of Toeplitz operators are always connected. Douglas's proof has subsequently been simplified by Searcóid [156].)

Exercise 3.4 is taken from [16], while Exercises 3.10, 3.12, 3.13, and 3.14 are extracted from Vukotić [168]. Cowen [76] contains Exercises 3.15, 3.16, and 3.17.

Another approach to Exercise 3.11 is suggested in Exercise 4.5 in Chapter 4 below. A necessary and sufficient condition that the "semicommutator" $T_{\phi\psi} - T_\phi T_\psi$ be compact (for general ϕ and ψ in $\boldsymbol{L}^\infty$) has been obtained by Axler, Chang, and Sarason [61], and Volberg [167].

Chapter 4

Hankel Operators

We introduce another interesting class of operators, the Hankel operators. Their matrices are obtained from the matrices of multiplication operators by taking a different corner from the one yielding Toeplitz matrices. We discuss some of the main properties of Hankel operators and their relationship to Toeplitz operators. We establish a number of results concerning boundedness, compactness, and spectral structure of Hankel operators.

4.1 Bounded Hankel Operators

Recall that the matrix of the multiplication operator M_ϕ, where $\phi \in \boldsymbol{L}^\infty$, is a Toeplitz matrix. That is,

$$M_\phi = \left(\begin{array}{cccc|cccc} \ddots & \ddots & \ddots & \ddots & \ddots & & & \\ \ddots & \phi_0 & \phi_{-1} & \phi_{-2} & \phi_{-3} & \ddots & & \\ \ddots & \phi_1 & \phi_0 & \phi_{-1} & \phi_{-2} & \phi_{-3} & \ddots & \\ \ddots & \phi_2 & \phi_1 & \phi_0 & \phi_{-1} & \phi_{-2} & \phi_{-3} & \ddots \\ \hline \ddots & \phi_3 & \phi_2 & \phi_1 & \boldsymbol{\phi_0} & \phi_{-1} & \phi_{-2} & \ddots \\ & \ddots & \phi_3 & \phi_2 & \phi_1 & \phi_0 & \phi_{-1} & \ddots \\ & & \ddots & \phi_3 & \phi_2 & \phi_1 & \phi_0 & \ddots \\ & & & \ddots & \ddots & \ddots & \ddots & \ddots \end{array}\right),$$

where ϕ_k is the kth Fourier coefficient of ϕ.

The matrices of Toeplitz operators are those that arise as lower-right corners of such matrices. We now consider upper-right corners. Each such corner is a matrix representing an operator from $\widetilde{\boldsymbol{H}}^2$ into $\boldsymbol{L}^2 \ominus \widetilde{\boldsymbol{H}}^2$. In order to get an operator mapping $\widetilde{\boldsymbol{H}}^2$ into itself, we need to apply the so-called flip operator.

Definition 4.1.1. The *flip operator* is the operator J mapping $\boldsymbol{L}^2$ into $\boldsymbol{L}^2$ defined by $(Jf)(e^{i\theta}) = f(e^{-i\theta})$.

It is clear that J is a unitary operator and that J is also self-adjoint. The matrix of JM_ϕ has the following form:

$$
JM_\phi = \left(\begin{array}{cccc|cccc}
\iddots & \iddots & \iddots & \iddots & \iddots & \iddots & & \\
\iddots & \phi_6 & \phi_5 & \phi_4 & \phi_3 & \iddots & \iddots & \\
\iddots & \phi_5 & \phi_4 & \phi_3 & \phi_2 & \phi_1 & \iddots & \iddots \\
\iddots & \phi_4 & \phi_3 & \phi_2 & \phi_1 & \phi_0 & \phi_{-1} & \iddots \\
\hline
\iddots & \phi_3 & \phi_2 & \phi_1 & \boldsymbol{\phi_0} & \phi_{-1} & \phi_{-2} & \iddots \\
\iddots & \iddots & \phi_1 & \phi_0 & \phi_{-1} & \phi_{-2} & \phi_{-3} & \iddots \\
 & \iddots & \iddots & \phi_{-1} & \phi_{-2} & \phi_{-3} & \phi_{-4} & \iddots \\
 & & \iddots & \iddots & \iddots & \iddots & \iddots & \iddots
\end{array} \right).
$$

This follows from the computation

$$
(JM_\phi e_n, e_m) = (M_\phi e^{in\theta}, e^{-im\theta}) = \frac{1}{2\pi} \int_0^{2\pi} \phi(e^{i\theta}) \overline{e^{-i(m+n)\theta}}\, d\theta = \phi_{-(m+n)}.
$$

The above matrix for JM_ϕ is constant along its skew-diagonals. Matrices of this form are known as Hankel matrices.

Definition 4.1.2. A finite matrix, or a doubly infinite matrix (i.e., a matrix with entries in positions (m, n) for m and n integers), or a singly infinite matrix (i.e., a matrix with entries in positions (m, n) for m and n nonnegative integers) is called a *Hankel matrix* if its entries are constant along each skew-diagonal. That is, the matrix $(a_{m,n})$ is Hankel if $a_{m_1,n_1} = a_{m_2,n_2}$ whenever $m_1 + n_1 = m_2 + n_2$.

Thus the matrix of JM_ϕ with respect to the standard basis of $\boldsymbol{L}^2$ is a doubly infinite Hankel matrix. It is clear that a bounded operator on $\boldsymbol{L}^2$ whose matrix with respect to the standard basis is a Hankel matrix is of the form JM_ϕ for some ϕ in $\boldsymbol{L}^\infty$. To see this, simply multiply the given Hankel matrix on the left by the matrix of J with respect to the standard basis. The resulting matrix is a doubly infinite Toeplitz matrix, and it is therefore the matrix of an operator M_ϕ with respect to the standard basis. Since $J^2 = I$ it follows that the original operator is JM_ϕ.

The study of singly infinite Hankel matrices is much more complicated than that of doubly infinite ones. Note that the lower-right corner of the matrix representation of JM_ϕ displayed above is a singly infinite Hankel matrix. That corner is the matrix of the restriction of PJM_ϕ to $\widetilde{\boldsymbol{H}}^2$ with respect to the standard basis of $\widetilde{\boldsymbol{H}}^2$.

Definition 4.1.3. A *Hankel operator* is an operator that is the restriction to $\widetilde{\boldsymbol{H}}^2$ of an operator of the form PJM_ϕ, where P is the projection of $\boldsymbol{L}^2$ onto $\widetilde{\boldsymbol{H}}^2$, J is the flip operator, and ϕ is a function in $\boldsymbol{L}^\infty$. This operator is denoted by H_ϕ. Its matrix with respect to the standard basis of $\widetilde{\boldsymbol{H}}^2$ is

$$\begin{pmatrix} \phi_0 & \phi_{-1} & \phi_{-2} & \phi_{-3} & \phi_{-4} & & \cdots \\ \phi_{-1} & \phi_{-2} & \phi_{-3} & \phi_{-4} & & \cdot^{\cdot^{\cdot}} & \\ \phi_{-2} & \phi_{-3} & \phi_{-4} & & \cdot^{\cdot^{\cdot}} & & \\ \phi_{-3} & \phi_{-4} & & \cdot^{\cdot^{\cdot}} & & & \\ \phi_{-4} & & \cdot^{\cdot^{\cdot}} & & & & \\ & \cdot^{\cdot^{\cdot}} & & & & & \\ \vdots & & & & & & \end{pmatrix},$$

where $\phi(e^{i\theta}) = \sum_{k=-\infty}^{\infty} \phi_k e^{ik\theta}$.

Note that every Hankel operator is bounded since it is the restriction of the product of three bounded operators.

As opposed to the situation with respect to Toeplitz operators, there is no unique symbol corresponding to a given Hankel operator.

Theorem 4.1.4. *The Hankel operators H_ϕ and H_ψ are equal if and only if $\phi - \psi$ is in $e^{i\theta}\widetilde{\boldsymbol{H}}^2$.*

Proof. Since the matrix of a Hankel operator depends only on the Fourier coefficients in nonpositive positions (Definition 4.1.3), two $\boldsymbol{L}^\infty$ functions induce the same Hankel operator if and only if their Fourier coefficients agree for nonpositive indices. This is equivalent to the difference between the functions being in $e^{i\theta}\widetilde{\boldsymbol{H}}^2$. $\square$

Definition 4.1.5. If f is a function in $\boldsymbol{L}^2$, then the *coanalytic part* of f is the function

$$\sum_{n=0}^{-\infty}(f, e^{in\theta})e^{in\theta}.$$

Thus two different functions with equal coanalytic parts induce the same Hankel operator, so we cannot talk about *the* symbol of a Hankel operator.

Definition 4.1.6. The $\boldsymbol{L}^\infty$ function ϕ is a *symbol* of the Hankel operator H if H is the restriction of PJM_ϕ to $\widetilde{\boldsymbol{H}}^2$.

Theorem 4.1.7. *The operator A has a Hankel matrix with respect to the standard basis of $\widetilde{\boldsymbol{H}}^2$ if and only if it satisfies the equation $U^*A = AU$, where U is the unilateral shift.*

Proof. First note that

$$(U^*Ae_n, e_m) = (Ae_n, Ue_m) = (Ae_n, e_{m+1}).$$

Also,

$$(AUe_n, e_m) = (Ae_{n+1}, e_m).$$

Therefore,

$$(U^*Ae_n, e_m) = (AUe_n, e_m)$$

for all m and n if and only if

$$(Ae_n, e_{m+1}) = (Ae_{n+1}, e_m)$$

for all m and n. Thus $U^*A = AU$ if and only if A has a Hankel matrix. □

Corollary 4.1.8. *If A has a Hankel matrix with respect to the standard basis of $\widetilde{\boldsymbol{H}}^2$, and U is the unilateral shift, then U^*AU has a Hankel matrix.*

Proof. This is easily seen by noticing the effect on the matrix of A of multiplying on the left by U^* and on the right by U. Alternatively,

$$\begin{aligned} U^*(U^*AU) &= U^*(U^*A)U \\ &= U^*(AU)U \quad \text{(by Theorem 4.1.7)} \\ &= (U^*AU)U. \end{aligned}$$

It follows from Theorem 4.1.7 that U^*AU is Hankel. □

It is not too easy to show that a bounded operator that has a Hankel matrix is a Hankel operator. To prove it requires several preliminary results.

Theorem 4.1.9 (Douglas's Theorem). *Let $\mathcal{H}$, $\mathcal{K}$, and $\mathcal{L}$ be Hilbert spaces and suppose that $E : \mathcal{H} \longrightarrow \mathcal{K}$ and $F : \mathcal{H} \longrightarrow \mathcal{L}$ are bounded operators. If $E^*E \leq F^*F$, then there exists an operator $R : \mathcal{L} \longrightarrow \mathcal{K}$ with $\|R\| \leq 1$ such that $E = RF$.*

Proof. First of all, observe that the hypothesis $E^*E \leq F^*F$ is equivalent to $\|Ex\| \leq \|Fx\|$ for all $x \in \mathcal{H}$, since $(E^*Ex, x) = \|Ex\|^2$ and $(F^*Fx, x) = \|Fx\|^2$.

We first define R on the range of F. Given Fx, set $RFx = Ex$. It needs to be shown that R is well-defined. If $Fx_1 = Fx_2$, then $F(x_1 - x_2) = 0$. Since $\|E(x_1 - x_2)\| \leq \|F(x_1 - x_2)\|$, we have $E(x_1 - x_2) = 0$ and hence $Ex_1 = Ex_2$. Thus $RFx_1 = RFx_2$.

That R is linear is obvious. We must show that R is bounded. In fact, $\|R\| \leq 1$, since $y = Fx$ yields

$$\|Ry\| = \|RFx\| = \|Ex\| \leq \|Fx\| = \|y\|.$$

Thus R is bounded, and we can extend it to the closure of the range of F by continuity. Define R to be zero on the orthogonal complement of the range of F. It is then clear that $\|R\| \leq 1$ and that $RFx = Ex$ for all $x \in \mathcal{H}$. □

Lemma 4.1.10. *Let $\mathcal{H}$ and $\mathcal{K}$ be Hilbert spaces and let $B : \mathcal{H} \longrightarrow \mathcal{K}$ be a bounded operator with $\|B\| \leq 1$. Then $(I - B^*B)^{1/2}B^* = B^*(I - BB^*)^{1/2}$.*

Proof. First of all, since $\|B\| = \|B^*\| \leq 1$, it follows that $I - B^*B \geq 0$ and $I - BB^* \geq 0$. Thus $(I - B^*B)^{1/2}$ and $(I - BB^*)^{1/2}$ exist, since every positive operator has a unique positive square root (see, for example, [12, p. 240], [48, p. 331]).

From the trivial equality $(I - B^*B)B^* = B^*(I - BB^*)$, it follows by induction that $(I - B^*B)^nB^* = B^*(I - BB^*)^n$ for every nonnegative integer n. By linearity, for every polynomial $p(x)$ we then have

$$p(I - B^*B)\, B^* = B^*\, p(I - BB^*).$$

By the Weierstrass approximation theorem [46, p. 159], there exists a sequence of polynomials $\{p_n\}$ such that $\{p_n(x)\}$ converges uniformly on $[0, 1]$ to $\sqrt{x}$. It follows that $\{p_n(I - B^*B)\}$ converges in norm to $(I - B^*B)^{1/2}$ and $\{p_n(I - BB^*)\}$ converges in norm to $(I - BB^*)^{1/2}$ (see, for example, [12, p.201], [42, pp. 222, 225]). Since

$$p_n(I - B^*B)\, B^* = B^*\, p_n(I - BB^*)$$

for all nonnegative integers n, taking limits gives

$$(I - B^*B)^{1/2}B = B^*(I - BB^*)^{1/2},$$

as desired. □

Theorem 4.1.11 (The Julia–Halmos Theorem). *Let $\mathcal{H}$ and $\mathcal{K}$ be Hilbert spaces and let $A : \mathcal{H} \longrightarrow \mathcal{K}$ be a bounded operator with $\|A\| \leq 1$. If U is the operator mapping $\mathcal{K} \oplus \mathcal{H}$ into $\mathcal{H} \oplus \mathcal{K}$ defined by the matrix*

$$U = \begin{pmatrix} -A^* & (I - A^*A)^{1/2} \\ (I - AA^*)^{1/2} & A \end{pmatrix},$$

then U is a unitary operator.

Proof. Applying Lemma 4.1.10 (with $B = A^*$ and with $B = A$) gives

$$(I - AA^*)^{1/2}A = A(I - A^*A)^{1/2} \quad \text{and} \quad (I - A^*A)^{1/2}A^* = A^*(I - AA^*)^{1/2}.$$

Multiplying matrices shows that the product, in both orders, of

$$\begin{pmatrix} -A & (I - AA^*)^{1/2} \\ (I - A^*A)^{1/2} & A^* \end{pmatrix}$$

and

$$\begin{pmatrix} -A^* & (I - A^*A)^{1/2} \\ (I - AA^*)^{1/2} & A \end{pmatrix}$$

is the identity. Hence U is unitary. □

Theorem 4.1.12 (Parrott's Theorem). *Let $\mathcal{H}_1$, $\mathcal{H}_2$, $\mathcal{K}_1$, and $\mathcal{K}_2$ be Hilbert spaces and $A : \mathcal{H}_2 \longrightarrow \mathcal{K}_2$, $B : \mathcal{H}_1 \longrightarrow \mathcal{K}_2$, and $C : \mathcal{H}_2 \longrightarrow \mathcal{K}_1$ be bounded operators. Then there exists a bounded operator $X : \mathcal{H}_1 \longrightarrow \mathcal{K}_1$ such that the operator*

$$\begin{pmatrix} X & C \\ B & A \end{pmatrix}$$

mapping $\mathcal{H}_1 \oplus \mathcal{H}_2$ into $\mathcal{K}_1 \oplus \mathcal{K}_2$ has norm equal to the maximum of

$$\left\| \begin{pmatrix} B & A \end{pmatrix} \right\| \quad \text{and} \quad \left\| \begin{pmatrix} C \\ A \end{pmatrix} \right\|.$$

Proof. For each operator X, let M_X denote the operator

$$\begin{pmatrix} X & C \\ B & A \end{pmatrix}.$$

We begin by proving that $\|M_X\|$ is at least as big as

$$\left\|\begin{pmatrix} B & A \end{pmatrix}\right\| \quad \text{and} \quad \left\|\begin{pmatrix} C \\ A \end{pmatrix}\right\|.$$

The first of these statements follows by noting that

$$\begin{pmatrix} B & A \end{pmatrix} = PM_X,$$

where P is the orthogonal projection from $\mathcal{K}_1 \oplus \mathcal{K}_2$ onto $\mathcal{K}_2$. Since $\|P\| = 1$,

$$\left\|\begin{pmatrix} B & A \end{pmatrix}\right\| = \|PM_X\| \le \|P\|\,\|M_X\| = \|M_X\|.$$

Similarly, the operator

$$\begin{pmatrix} C \\ A \end{pmatrix}$$

is the restriction of M_X to $\mathcal{H}_2$, and thus clearly has norm at most that of M_X.

Conversely, first note that we can take $X = 0$ in the case

$$\left\|\begin{pmatrix} B & A \end{pmatrix}\right\| = \left\|\begin{pmatrix} C \\ A \end{pmatrix}\right\| = 0.$$

In the other cases, dividing A, B, and C all by the maximum of

$$\left\|\begin{pmatrix} B & A \end{pmatrix}\right\| \quad \text{and} \quad \left\|\begin{pmatrix} C \\ A \end{pmatrix}\right\|$$

allows us to assume that that maximum is 1.

Thus we may, and do, suppose that

$$\left\|\begin{pmatrix} B & A \end{pmatrix}\right\| \le 1 \quad \text{and} \quad \left\|\begin{pmatrix} C \\ A \end{pmatrix}\right\| \le 1$$

and at least one of these inequalities is an equality.

Note that

$$\begin{pmatrix} B^* \\ A^* \end{pmatrix} = \begin{pmatrix} B & A \end{pmatrix}^*$$

and

$$\begin{pmatrix} C \\ A \end{pmatrix} = \begin{pmatrix} C^* & A^* \end{pmatrix}^* .$$

Recall (as is very easily verified) that $I - T^*T \geq 0$ when the operator T has norm at most 1. Hence,

$$I - \begin{pmatrix} B & A \end{pmatrix} \begin{pmatrix} B^* \\ A* \end{pmatrix} \geq 0 \quad \text{and} \quad I - \begin{pmatrix} C^* & A^* \end{pmatrix} \begin{pmatrix} C \\ A \end{pmatrix} \geq 0,$$

or, equivalently,

$$I - BB^* - AA^* \geq 0 \quad \text{and} \quad I - C^*C - A^*A \geq 0.$$

Observe that the above expressions are equivalent to

$$BB^* \leq I - AA^* \quad \text{and} \quad C^*C \leq I - A^*A.$$

Since, by the above inequalities, $I - AA^*$ and $I - A^*A$ are positive operators, they each have unique positive square roots (e.g., [12, p. 240], [48, p. 331]). Thus

$$BB^* \leq (I - AA^*)^{1/2}(I - AA^*)^{1/2} \quad \text{and} \quad C^*C \leq (I - A^*A)^{1/2}(I - A^*A)^{1/2}.$$

Douglas's theorem (Theorem 4.1.9) implies that there exist operators T and R of norm at most 1 such that

$$B^* = T(I - AA^*)^{1/2} \quad \text{and} \quad C = R(I - A^*A)^{1/2}.$$

Straightforward matrix multiplication yields

$$\begin{pmatrix} R & 0 \\ 0 & I \end{pmatrix} \begin{pmatrix} -A^* & (I - A^*A)^{1/2} \\ (I - AA^*)^{1/2} & A \end{pmatrix} \begin{pmatrix} T^* & 0 \\ 0 & I \end{pmatrix} = \begin{pmatrix} -RA^*T^* & C \\ B & A \end{pmatrix}.$$

Note that

$$\left\| \begin{pmatrix} B \\ A \end{pmatrix} \right\| \leq 1$$

implies $\|A\| \leq 1$; hence the Julia–Halmos theorem (Theorem 4.1.11) gives, in particular,

$$\left\| \begin{pmatrix} -A^* & (I - A^*A)^{1/2} \\ (I - AA^*)^{1/2} & A \end{pmatrix} \right\| = 1.$$

Since T and R have norm less than or equal to 1, it follows that

$$\left\| \begin{pmatrix} R & 0 \\ 0 & I \end{pmatrix} \right\| \leq 1 \quad \text{and} \quad \left\| \begin{pmatrix} T^* & 0 \\ 0 & I \end{pmatrix} \right\| \leq 1.$$

Thus the above equation gives

$$\left\| \begin{pmatrix} -RA^*T^* & C \\ B & A \end{pmatrix} \right\| \leq 1.$$

Therefore taking $X = -RA^*T^*$ gives the result. □

We can now show that a bounded operator that has a Hankel matrix is a Hankel operator.

Theorem 4.1.13 (Nehari's Theorem). *A Hankel matrix corresponds to a bounded operator H on $\widetilde{\boldsymbol{H}}^2$ if and only if there exists a function ϕ in $\boldsymbol{L}^\infty$ such that $H = H_\phi = PJM_\phi \big|_{\widetilde{\boldsymbol{H}}^2}$. Moreover, ϕ can be chosen such that $\|H_\phi\| = \|\phi\|_\infty$.*

Proof. We already showed (see the comment before Definition 4.1.3) that the matrix with respect to the standard basis of the restriction of PJM_ϕ to $\widetilde{\boldsymbol{H}}^2$ is a Hankel matrix.

Conversely, suppose that H is a bounded operator whose matrix with respect to the standard basis of $\widetilde{\boldsymbol{H}}^2$ is the Hankel matrix

$$\begin{pmatrix} a_0 & a_{-1} & a_{-2} & a_{-3} & a_{-4} & \cdots \\ a_{-1} & a_{-2} & a_{-3} & a_{-4} & \cdot^{\cdot^{\cdot}} & \\ a_{-2} & a_{-3} & a_{-4} & \cdot^{\cdot^{\cdot}} & & \\ a_{-3} & a_{-4} & \cdot^{\cdot^{\cdot}} & & & \\ a_{-4} & \cdot^{\cdot^{\cdot}} & & & & \\ \vdots & & & & & \end{pmatrix}.$$

We must find a function ϕ in $\boldsymbol{L}^\infty$ whose Fourier coefficients with nonpositive indices are the $\{a_{-k}\}$; that is, a function $\phi \in \boldsymbol{L}^\infty$ such that

$$\frac{1}{2\pi}\int_0^{2\pi} \phi(e^{i\theta})e^{ik\theta}\,d\theta = a_{-k} \quad \text{for } k = 0, 1, 2, \ldots .$$

It is clear that there are many such functions ϕ in $\boldsymbol{L}^2$. We must use the fact that the above matrix represents a bounded operator to establish that there is such a ϕ in $\boldsymbol{L}^\infty$.

We proceed as follows. We will inductively construct coefficients $a_1, a_2, a_3, \ldots$ such that the doubly infinite Hankel matrix

$$
\left(\begin{array}{cccc|cccc}
\iddots & \iddots & \iddots & \iddots & \iddots & \iddots & & \\
\iddots & a_6 & a_5 & a_4 & a_3 & \iddots & \iddots & \\
\iddots & a_5 & a_4 & a_3 & a_2 & a_1 & \iddots & \iddots \\
\iddots & a_4 & a_3 & a_2 & a_1 & a_0 & a_{-1} & \iddots \\
\hline
\iddots & a_3 & a_2 & a_1 & a_0 & a_{-1} & a_{-2} & \iddots \\
\iddots & \iddots & a_1 & a_0 & a_{-1} & a_{-2} & a_{-3} & \iddots \\
 & \iddots & \iddots & a_{-1} & a_{-2} & a_{-3} & a_{-4} & \iddots \\
 & & \iddots & \iddots & \iddots & \iddots & \iddots & \iddots
\end{array}\right)
$$

is bounded and has the same norm as H.

Consider the matrix

$$
M_x = \left(\begin{array}{c|ccccc}
x & a_0 & a_{-1} & a_{-2} & a_{-3} & \cdots \\
\hline
a_0 & a_{-1} & a_{-2} & a_{-3} & & \iddots \\
a_{-1} & a_{-2} & a_{-3} & & \iddots & \\
a_{-2} & a_{-3} & & \iddots & & \\
a_{-3} & & \iddots & & & \\
 & \iddots & & & & \\
\vdots & & & & &
\end{array}\right),
$$

and think of it as the 2×2 operator matrix

$$
M_x = \left(\begin{array}{c|c} x & C \\ \hline B & A \end{array}\right).
$$

Apply Parrott's theorem (Theorem 4.1.12) to see that there exists a complex number a_1 such that the norm of

$$
\begin{pmatrix}
a_1 & a_0 & a_{-1} & a_{-2} & a_{-3} & \cdots \\
a_0 & a_{-1} & a_{-2} & a_{-3} & \iddots & \\
a_{-1} & a_{-2} & a_{-3} & \iddots & & \\
a_{-2} & a_{-3} & \iddots & & & \\
a_{-3} & \iddots & & & & \\
\vdots & & & & &
\end{pmatrix}
$$

is equal to the norm of H.

For the second step in the induction, we must show that there is an a_2 such that the norm of the matrix

$$\begin{pmatrix} a_2 & a_1 & a_0 & a_{-1} & a_{-2} & a_{-3} & \cdots \\ a_1 & a_0 & a_{-1} & a_{-2} & a_{-3} & \cdot^{\cdot^{\cdot}} & \\ a_0 & a_{-1} & a_{-2} & a_{-3} & \cdot^{\cdot^{\cdot}} & & \\ a_{-1} & a_{-2} & a_{-3} & \cdot^{\cdot^{\cdot}} & & & \\ a_{-2} & a_{-3} & \cdot^{\cdot^{\cdot}} & & & & \\ a_{-3} & \cdot^{\cdot^{\cdot}} & & & & & \\ \vdots & & & & & & \end{pmatrix}$$

is equal to the norm of H. This also follows from Parrott's theorem (Theorem 4.1.12), with the operators

$$\begin{pmatrix} B & A \end{pmatrix} \quad \text{and} \quad \begin{pmatrix} C \\ A \end{pmatrix}$$

both equal to

$$\begin{pmatrix} a_1 & a_0 & a_{-1} & a_{-2} & a_{-3} & \cdots \\ a_0 & a_{-1} & a_{-2} & a_{-3} & \cdot^{\cdot^{\cdot}} & \\ a_{-1} & a_{-2} & a_{-3} & \cdot^{\cdot^{\cdot}} & & \\ a_{-2} & a_{-3} & \cdot^{\cdot^{\cdot}} & & & \\ a_{-3} & \cdot^{\cdot^{\cdot}} & & & & \\ \vdots & & & & & \end{pmatrix}.$$

Inductively, suppose we have found complex numbers $a_1, a_2, a_3, \ldots, a_k$ such that, for each $s = 1, 2, \ldots, k$, the matrix

$$\begin{pmatrix}
a_s & a_{s-1} & \cdots\cdots & a_1 & a_0 & a_{-1} & a_{-2} & & \cdots \\
a_{s-1} & & a_1 & a_0 & a_{-1} & a_{-2} & & \iddots & \\
\vdots & & a_1 & a_0 & a-1 & a_{-2} & \iddots & & \\
\vdots & a_1 & a_0 & a_{-1} & a_{-2} & \iddots & & & \\
a_1 & a_0 & a_{-1} & a_{-2} & \iddots & & & & \\
a_0 & a_{-1} & a_{-2} & \iddots & & & & & \\
a_{-1} & a_{-2} & \iddots & & & & & & \\
a_{-2} & \iddots & & & & & & & \\
\iddots & & & & & & & & \\
\vdots & & & & & & & &
\end{pmatrix}$$

has norm equal to $\|H\|$. Using Parrott's theorem as above, we can find a_{k+1} such that

$$\begin{pmatrix}
a_{k+1} & a_k & \cdots\;\cdots & a_1 & a_0 & a_{-1} & a_{-2} & & \cdots \\
a_k & & a_1 & a_0 & a_{-1} & a_{-2} & & \iddots & \\
\vdots & & a_1 & a_0 & a-1 & a_{-2} & \iddots & & \\
\vdots & a_1 & a_0 & a_{-1} & a_{-2} & \iddots & & & \\
a_1 & a_0 & a_{-1} & a_{-2} & \iddots & & & & \\
a_0 & a_{-1} & a_{-2} & \iddots & & & & & \\
a_{-1} & a_{-2} & \iddots & & & & & & \\
a_{-2} & \iddots & & & & & & & \\
\iddots & & & & & & & & \\
\vdots & & & & & & & &
\end{pmatrix}$$

has norm equal to $\|H\|$.

Consider the doubly infinite Hankel matrix Γ given by

$$\Gamma = \left(\begin{array}{cccc|cccc}
\iddots & \iddots & \iddots & \iddots & \iddots & \iddots & & \\
\iddots & a_6 & a_5 & a_4 & a_3 & \iddots & \iddots & \\
\iddots & a_5 & a_4 & a_3 & a_2 & a_1 & \iddots & \iddots \\
\iddots & a_4 & a_3 & a_2 & a_1 & a_0 & a_{-1} & \iddots \\
\hline
\iddots & a_3 & a_2 & a_1 & \mathbf{a_0} & a_{-1} & a_{-2} & \iddots \\
\iddots & \iddots & a_1 & a_0 & a_{-1} & a_{-2} & a_{-3} & \iddots \\
 & \iddots & \iddots & a_{-1} & a_{-2} & a_{-3} & a_{-4} & \iddots \\
 & & \iddots & \iddots & \iddots & \iddots & \iddots & \iddots
\end{array}\right).$$

Note that, for each natural number n, the section of Γ defined by

$$\Gamma_n = \begin{pmatrix} a_n & a_{n-1} & \cdots & \cdots & a_1 & a_0 & a_{-1} & a_{-2} & \cdots \\ a_{n-1} & & & a_1 & a_0 & a_{-1} & a_{-2} & \iddots & \\ \vdots & & a_1 & a_0 & a-1 & a_{-2} & \iddots & & \\ \vdots & a_1 & a_0 & a_{-1} & a_{-2} & \iddots & & & \\ a_1 & a_0 & a_{-1} & a_{-2} & \iddots & & & & \\ a_0 & a_{-1} & a_{-2} & \iddots & & & & & \\ a_{-1} & a_{-2} & \iddots & & & & & & \\ a_{-2} & \iddots & & & & & & & \\ \vdots & & & & & & & & \end{pmatrix}$$

has norm equal to $\|H\|$.

Now, Γ is the matrix of an operator with respect to a basis $\{e_j\}_{j=-\infty}^{\infty}$, where $\bigvee\{e_j : j \geq 0\} = \widetilde{\boldsymbol{H}}^2$. For each negative integer N, let P_N denote the projection onto the span of $\bigvee\{e_j : j \geq N\}$. For every vector x, it is clear that $\|\Gamma x\|$ is the supremum of $\|P_N \Gamma P_N x\|$. Therefore, $\|\Gamma x\| \leq \|H\| \|x\|$ for every x, so $\|\Gamma\| \leq \|H\|$. On the other hand, for vectors x in $\widetilde{\boldsymbol{H}}^2$, $\|\Gamma x\| \geq \|Hx\|$, so $\|\Gamma\| \geq \|H\|$.

Therefore Γ is the matrix of a bounded operator whose norm is $\|H\|$.

Let J denote the flip operator (see Definition 4.1.1). Then the matrix of $J\Gamma$ is given by

$$J\Gamma = \left(\begin{array}{cccc|cccc} \iddots & \iddots & \iddots & \iddots & \iddots & \iddots & & \\ \iddots & a_0 & a_{-1} & a_{-2} & a_{-3} & \iddots & \iddots & \\ \iddots & a_1 & a_0 & a_{-1} & a_{-2} & a_{-3} & \iddots & \iddots \\ \iddots & a_2 & a_1 & a_0 & a_{-1} & a_{-2} & a_{-3} & \iddots \\ \hline \iddots & a_3 & a_2 & a_1 & a_0 & a_{-1} & a_{-2} & \iddots \\ \iddots & \iddots & a_3 & a_2 & a_1 & a_0 & a_{-1} & \iddots \\ & \iddots & \iddots & a_3 & a_2 & a_1 & a_0 & \iddots \\ & & \iddots & \iddots & \iddots & \iddots & \iddots & \iddots \end{array}\right).$$

Thus $J\Gamma$ is a doubly infinite Toeplitz matrix of a bounded operator, and, by Theorem 3.1.4, there exists a function $\phi \in \boldsymbol{L}^\infty$ such that $J\Gamma = M_\phi$ and $\|M_\phi\| = \|\phi\|_\infty$. Since $J^2 = I$, $\Gamma = JM_\phi$, and, since J is unitary, $\|\Gamma\| =$

$\|M_\phi\| = \|\phi\|_\infty$. Restricting to $\widetilde{\boldsymbol{H}}^2$ and multiplying on the left by P gives

$$H = PJM_\phi \Big|_{\widetilde{\boldsymbol{H}}^2} .$$

Clearly, $\|H\| = \|\Gamma\| = \|\phi\|_\infty$. □

4.2 Hankel Operators of Finite Rank

There are many Hankel operators of finite rank. For example, if the function ϕ in $\boldsymbol{L}^\infty$ has only a finite number of nonzero Fourier coefficients in positions of negative index, then clearly H_ϕ has finite rank. In fact, in those cases, the standard matrix representation of H_ϕ has only a finite number of entries other than zero.

There are also finite-rank Hankel operators whose standard matrices contain an infinite number of entries other than zero. As will be shown, the Hankel operators of rank 1 arise from the kernel functions for $\boldsymbol{H}^2$.

We will find it useful to have notation for the function obtained from a given function by interchanging its analytic and coanalytic parts.

Notation 4.2.1. For each function f in $\boldsymbol{L}^2$, the function $\breve{f}$ is defined by

$$\breve{f}(e^{i\theta}) = f(e^{-i\theta}).$$

Obviously, $\|f\| = \|\breve{f}\|$ and, if f is in $\boldsymbol{L}^\infty$, $\|f\|_\infty = \|\breve{f}\|_\infty$.

Note that $Jf = \breve{f}$, where J is the flip operator (Definition 4.1.1).

Theorem 4.2.2. *For a fixed element w of $\mathbb{D}$, let k_w be defined by*

$$k_w(z) = \frac{1}{1 - \overline{w}z}$$

(see Definition 1.1.7). Then the rank-one operator $k_{\overline{w}} \otimes k_w$ (see Notation 1.2.27) is the Hankel operator $H_{\breve{k}_{\overline{w}}}$. Its matrix with respect to the standard basis of $\widetilde{\boldsymbol{H}}^2$ is

$$\begin{pmatrix} 1 & w & w^2 & w^3 & & \cdots \\ w & w^2 & w^3 & & \iddots & \\ w^2 & w^3 & & \iddots & & \\ w^3 & & \iddots & & & \\ & \iddots & & & & \\ \vdots & & & & & \\ & & & & & \end{pmatrix}.$$

Conversely, if H is a Hankel operator whose rank is 1, then there exists a w in $\mathbb{D}$ and a constant c such that $H = c(k_{\overline{w}} \otimes k_w)$.

Proof. To see that $k_{\overline{w}} \otimes k_w$ has the stated matrix representation, fix any nonnegative integer n and compute

$$(k_{\overline{w}} \otimes k_w)\, e_n = (e_n, k_w)\, k_{\overline{w}} = w^n\, (e_0 + we_1 + w^2 e_2 + w^3 e_3 + \cdots).$$

Hence the matrix of $k_{\overline{w}} \otimes k_w$ is

$$\begin{pmatrix} 1 & w & w^2 & w^3 & \cdots \\ w & w^2 & w^3 & \cdot^{\cdot^{\cdot}} & \\ w^2 & w^3 & \cdot^{\cdot^{\cdot}} & & \\ w^3 & \cdot^{\cdot^{\cdot}} & & & \\ & \cdot^{\cdot^{\cdot}} & & & \\ \vdots & & & & \end{pmatrix}.$$

Clearly, the above matrix is the matrix of the Hankel operator having symbol $\breve{k}_{\overline{w}}$.

To establish the converse, that every Hankel operator of rank 1 is a multiple of an operator of the given form, proceed as follows. Let $f \otimes g$ be an arbitrary rank-one operator in $\widetilde{\boldsymbol{H}}^2$. Suppose that $f \otimes g$ is Hankel. Then

$$U^*(f \otimes g) = (f \otimes g)U \qquad \text{(by Theorem 4.1.7)}$$

so

$$(U^* f) \otimes g = f \otimes (U^* g) \qquad \text{(by Theorem 1.2.28).}$$

Since the equality of rank-one operators implies that their defining vectors are multiples of each other (Theorem 1.2.28), it follows that there exists a constant w such that $U^* f = wf$ and $U^* g = \overline{w} g$. Thus w is an eigenvalue of the backward unilateral shift, so $|w| < 1$ (Theorem 2.1.6). The eigenvectors of U^* corresponding to the eigenvalue w are all multiples of $k_{\overline{w}}$ (Theorem 2.1.6), so there are constants c_1 and c_2 such that $f = c_1 k_{\overline{w}}$ and $g = c_2 k_w$. Thus $f \otimes g = c_1 c_2 (k_{\overline{w}} \otimes k_w)$. □

It is clear that linear combinations of Hankel operators are Hankel operators, so

$$H_{\breve{k}_{\overline{w_1}}} + H_{\breve{k}_{\overline{w_2}}}$$

is a Hankel operator of rank 2 whenever w_1 and w_2 are distinct points of $\mathbb{D}$. It is interesting to note that there are Hankel operators of rank 2 that cannot be written as the sum of Hankel operators of rank one (see Exercise 4.11).

Note that the function $\breve{k}_{\overline{w}}$ is defined on S^1 by

$$\begin{aligned}\breve{k}_{\overline{w}}(e^{i\theta}) &= k_{\overline{w}}(e^{-i\theta}) \\ &= \frac{1}{1-we^{-i\theta}} \\ &= 1 + we^{-i\theta} + w^2e^{-2i\theta} + w^3e^{-3i\theta} + \cdots \\ &= \sum_{n=0}^{\infty} \left(\frac{w}{e^{i\theta}}\right)^n.\end{aligned}$$

If $|z| > |w|$, the series $\sum_{n=0}^{\infty} \left(\frac{w}{z}\right)^n$ converges to

$$\frac{1}{1-\frac{w}{z}} = \frac{z}{z-w}.$$

Thus $\breve{k}_{\overline{w}}$ has the meromorphic extension

$$\breve{k}_{\overline{w}}(z) = \frac{z}{z-w}$$

to the plane.

We shall show that a Hankel operator has finite rank if and only if it has a symbol that has an extension to a rational function on $\mathbb{C}$.

More generally, we establish the following necessary and sufficient condition that the columns of a Hankel matrix (whether or not it is the matrix of a bounded operator) span a finite dimensional space.

Theorem 4.2.3 (Kronecker's Theorem). *Let*

$$H = \begin{pmatrix} a_0 & a_1 & a_2 & a_3 & & \cdots \\ a_1 & a_2 & a_3 & & \cdot^{\cdot^{\cdot}} & \\ a_2 & a_3 & & \cdot^{\cdot^{\cdot}} & & \\ a_3 & & \cdot^{\cdot^{\cdot}} & & & \\ & \cdot^{\cdot^{\cdot}} & & & & \\ \vdots & & & & & \end{pmatrix}$$

be any singly infinite Hankel matrix. If the columns of H are linearly dependent (in the space of all sequences), then

$$a_0 + a_1/z + a_2/z^2 + a_3/z^3 + \cdots$$

is a rational function. In the other direction, if

$$a_0 + a_1/z + a_2/z^2 + a_3/z^3 + \cdots$$

is a rational function, then the columns of H span a finite-dimensional subspace of the space of all sequences.

Proof. Suppose that the columns of H are linearly dependent. Then there is a natural number s and complex numbers $c_0, c_1, c_2, \ldots, c_s$, not all zero, such that

$$c_0 \begin{pmatrix} a_0 \\ a_1 \\ a_2 \\ a_3 \\ \vdots \end{pmatrix} + c_1 \begin{pmatrix} a_1 \\ a_2 \\ a_3 \\ a_4 \\ \vdots \end{pmatrix} + c_2 \begin{pmatrix} a_2 \\ a_3 \\ a_4 \\ a_5 \\ \vdots \end{pmatrix} + \cdots + c_s \begin{pmatrix} a_s \\ a_{s+1} \\ a_{s+2} \\ a_{s+3} \\ \vdots \end{pmatrix} = \begin{pmatrix} 0 \\ 0 \\ 0 \\ 0 \\ \vdots \end{pmatrix}.$$

This means that

$$\sum_{j=0}^{s} c_j a_{l+j} = 0,$$

for all nonnegative integers l.

Let p denote the polynomial $p(z) = c_0 + c_1 z + c_2 z^2 + \cdots + c_s z^s$ and let f denote the function $f(z) = a_0 + a_1/z + a_2/z^2 + a_3/z^3 + \cdots$. To show that f is a rational function, it suffices to show that pf is a polynomial. We compute

$$p(z)f(z) = \sum_{j=0}^{s} c_j z^j \sum_{k=0}^{\infty} a_k z^{-k} = \sum_{j=0}^{s} \sum_{k=0}^{\infty} c_j a_k z^{j-k}.$$

Define the polynomial q by

$$q(z) = \sum_{j=0}^{s} \sum_{k=0}^{j} c_j a_k z^{j-k};$$

i.e., q is the sum of the terms for which $j - k \geq 0$ in the double sum. Note that

$$\sum_{j=0}^{s} \sum_{k=j+1}^{\infty} c_j a_k z^{j-k} = \sum_{l=1}^{\infty} \sum_{j=0}^{s} c_j a_{j+l} z^{-l},$$

as can be seen by letting $l = k - j$ and interchanging the order of summation. However, as shown above, the assumed linear dependence implies that

$$\sum_{j=0}^{s} c_j a_{j+l} = 0$$

for every l. Therefore, $p(z)f(z) = q(z)$, so $f(z) = q(z)/p(z)$ is a rational function, as desired.

The proof of the other direction of the theorem is similar to the above. Assume that

$$f(z) = \sum_{k=0}^{\infty} a_k z^{-k}$$

is a rational function. Then there exist polynomials p and q such that $f(z) = q(z)/p(z)$. Note that the degree of q is at most the degree of p, since $f(z)$ has no polynomial part.

Suppose that

$$p(z) = c_0 + c_1 z + c_2 z^2 + \cdots + c_s z^s.$$

Then

$$p(z)f(z) = \sum_{j=0}^{s} c_j z^j \sum_{k=0}^{\infty} a_k z^{-k} = \sum_{j=0}^{s} \sum_{k=0}^{\infty} c_j a_k z^{j-k} = q(z).$$

Thus

$$\sum_{j=0}^{s} \sum_{k=0}^{\infty} c_j a_k z^{j-k}$$

is a polynomial of degree at most s. Making the change of variables $l = k - j$,

$$\begin{aligned} \sum_{j=0}^{s} \sum_{k=0}^{\infty} c_j a_k z^{j-k} &= \sum_{j=0}^{s} \sum_{l=-j}^{\infty} c_j a_{j+l} z^{-l} \\ &= \sum_{j=0}^{s} \sum_{l=-j}^{0} c_j a_{j+l} z^{-l} + \sum_{j=0}^{s} \sum_{l=1}^{\infty} c_j a_{j+l} z^{-l}. \end{aligned}$$

Since this is a polynomial, it follows that

$$\sum_{j=0}^{s} \sum_{l=1}^{\infty} c_j a_{j+l} z^{-l} = \sum_{l=1}^{\infty} \left(\sum_{j=0}^{s} c_j a_{j+l} \right) z^{-l} = 0.$$

Hence

$$\sum_{j=0}^{s} c_j a_{j+l} = 0$$

for every natural number l.

The above collection of equations for natural numbers l implies that every family of $s + 1$ consecutive columns of H that does not include the first column is linearly dependent. Hence every column is in the span of the first $s + 2$ columns of H, so the span of all the columns of H has dimension at most $s + 2$. □

In the case that the Hankel matrix represents a bounded operator, the above theorem can be slightly restated; the other formulation is also frequently referred to as "Kronecker's theorem". Note that a rational function with no poles on the unit circle is continuous on the circle, and is therefore in $\boldsymbol{L}^\infty$.

Recall that a Hankel operator depends only on the coanalytic part of its symbol (Definition 4.1.5).

Corollary 4.2.4 (Kronecker's Theorem). *For $\phi \in \boldsymbol{L}^\infty$, the Hankel operator H_ϕ has finite rank if and only if the function*

$$f(z) = \sum_{k=0}^{\infty} \frac{\phi_{-k}}{z^k}$$

is a rational function all of whose poles are in $\mathbb{D}$.

Proof. Suppose H_ϕ has finite rank. Then f is a rational function, by the version of Kronecker's theorem we have already established (Theorem 4.2.3). It must be shown that f has no poles outside $\mathbb{D}$.

If f had a pole outside $\mathbb{D}$, say λ, consider the rational function g defined as $g(z) = f(1/z)$. If we consider the function $H_\phi(e_0)$ as a function in $\boldsymbol{H}^2$, it is easy to see that $H_\phi(e_0) = g$. But this cannot occur because g cannot be in $\boldsymbol{H}^2$. Indeed, if $|\lambda| > 1$, then the function g has a pole at $1/\lambda \in \mathbb{D}$ and hence cannot be analytic. If $|\lambda| = 1$ then g cannot be in $\boldsymbol{H}^2$ since rational functions with poles on S^1 cannot be in $\boldsymbol{H}^2$ (see Exercise 4.7). Therefore f must have all its poles in $\mathbb{D}$.

Conversely, assume that f is a rational function with all its poles in $\mathbb{D}$. Then f is continuous on S^1 and hence it is in $\boldsymbol{L}^\infty$. By the previous version of Kronecker's theorem (Theorem 4.2.3), H_ϕ has finite rank. But clearly $H_f = H_\phi$. □

A slightly different formulation of Kronecker's theorem in the case of bounded operators is the following.

Corollary 4.2.5 (Kronecker's Theorem). *Let $\mathcal{R}$ be the set of rational functions with poles inside $\mathbb{D}$. Then H is a bounded Hankel operator of finite rank if and only if $H = H_\psi$ for some $\psi \in e^{i\theta}\widetilde{\boldsymbol{H}^\infty} + \mathcal{R}$.*

Proof. The previous form of Kronecker's theorem (Corollary 4.2.4) immediately implies this form since Hankel operators are the same if and only if their symbols have identical coanalytic parts (Theorem 4.1.4). □

The Hankel operators whose matrices with respect to the standard basis for $\widetilde{\boldsymbol{H}}^2$ have only a finite number of entries different from 0 can be described very explicitly. Note that H_ϕ has this property whenever $\phi = e^{-in\theta}\widetilde{\psi}$ for any $\psi \in \boldsymbol{H}^2$. It is also obvious that if H_ϕ has this property, then $e^{in\theta}\phi$ is in $\widetilde{\boldsymbol{H}}^2$ for some nonnegative integer n. The following sharpening of this statement is less obvious.

Theorem 4.2.6. *If a Hankel operator has a matrix with respect to the standard basis for $\widetilde{\boldsymbol{H}}^2$ that has only a finite number of entries different from 0, and if c is the norm of the operator, then there exists a finite Blaschke product B and a nonnegative integer n such that $ce^{-in\theta}\widetilde{B}$ is a symbol for the Hankel operator.*

Proof. The theorem is trivial if the operator is 0; in all other cases, divide the operator by its norm to reduce to the case in which the norm is 1.

By Nehari's theorem (Theorem 4.1.13), the Hankel operator has a symbol ψ in $\boldsymbol{L}^\infty$ such that $\|H_\psi\| = \|\psi\|_\infty = 1$. We shall show that the function ψ has the desired form.

Since the coanalytic part of ψ has finitely many nonzero coefficients, Kronecker's theorem (Corollary 4.2.4) implies that H_ψ is an operator of finite rank. Since every finite-rank operator achieves its norm, there is a unit vector g in $\widetilde{\boldsymbol{H}}^2$ such that

$$\|H_\psi g\| = \|H_\psi\|.$$

We then have

$$1 = \|H_\psi\| = \|H_\psi g\| = \|PJM_\psi g\| = \|P\breve{\psi}\breve{g}\| \leq \|\breve{\psi}\breve{g}\| \leq \|\breve{\psi}\|_\infty \|\breve{g}\| = 1,$$

and thus the inequalities above become equalities. In particular, we have

$$\|P\breve{\psi}\breve{g}\| = \|\breve{\psi}\breve{g}\| \quad \text{and} \quad \|\breve{\psi}\breve{g}\| = 1.$$

Since $\|P\breve{\psi}\breve{g}\| = \|\breve{\psi}\breve{g}\|$, the function $\breve{\psi}\breve{g}$ is in $\widetilde{\boldsymbol{H}}^2$ and hence the function ψg is in $(e^{i\theta}\widetilde{\boldsymbol{H}}^2)^\perp$.

Since the matrix of H_ψ has only a finite number of entries different from 0, there is a nonnegative integer n such that $e^{in\theta}\psi$ is in $\widetilde{\boldsymbol{H}}^\infty$. It follows that the function $e^{in\theta}\psi g$ is in $\widetilde{\boldsymbol{H}}^2$ and hence ψg is in $e^{-in\theta}\widetilde{\boldsymbol{H}}^2$. But the only way that ψg can be in both of $(e^{i\theta}\widetilde{\boldsymbol{H}}^2)^\perp$ and $e^{-in\theta}\widetilde{\boldsymbol{H}}^2$ is if ψg is a polynomial in $e^{-i\theta}$ of degree at most n. Therefore $e^{in\theta}\psi g$ is a polynomial p in $e^{i\theta}$ of degree at most n.

Recall that $\|\breve{\psi}\breve{g}\| = 1$. Since $\|\breve{g}\| = 1$ and $|\breve{\psi}(e^{i\theta})| \leq 1$ a.e., we must have $|\breve{\psi}(e^{i\theta})| = 1$ a.e. Hence $|e^{in\theta}\psi(e^{i\theta})| = 1$ a.e. as well. Since $e^{in\theta}\psi$ is in $\widetilde{\boldsymbol{H}}^2$, it then follows that $e^{in\theta}\psi$ is an inner function. But since $e^{in\theta}\psi g = p$ is a polynomial, this implies that the inner function $e^{in\theta}\psi$ divides the inner factor of p (Theorem 2.6.7).

Since a polynomial is analytic on the entire plane and has only a finite number of zeros, the inner part of a polynomial must be a finite Blaschke product. Therefore $e^{in\theta}\psi$ is a finite Blaschke product, which proves the theorem. □

Note that the $\boldsymbol{L}^\infty$-norm of the symbol obtained in the previous theorem is the norm of the Hankel operator, since $|e^{-in\theta}|$ equals 1 for all θ.

We need the following observation.

Corollary 4.2.7. *If the matrix of a Hankel operator with respect to the standard basis for $\widetilde{\boldsymbol{H}}^2$ has only a finite number of entries different from 0, then there exists a function ψ, continuous on S^1, that is a symbol of the Hankel operator and satisfies $\|H_\psi\| = \|\psi\|_\infty$.*

Proof. Since any finite Blaschke product is continuous on S^1, the result follows from the previous theorem. □

4.3 Compact Hankel Operators

The natural question of when a Hankel operator is compact arises. The answer is given by the classical theorem of Hartman.

Theorem 4.3.1 (Hartman's Theorem). *A Hankel operator H is compact if and only if there exists a continuous function ϕ such that $H = H_\phi$. Every compact Hankel operator is a uniform limit of Hankel operators of finite rank.*

Proof. First assume that ϕ is continuous on S^1. The fact that H_ϕ is compact follows very easily from the Weierstrass approximation theorem [47, p. 91]. To see this, simply choose a sequence $\{p_n\}$ that converges uniformly to ϕ on S^1. By Kronecker's theorem (Corollary 4.2.4), the Hankel operator with symbol p_n has finite rank. But then

$$\|H_{p_n} - H_\phi\| = \|H_{p_n - \phi}\| \leq \|p_n - \phi\|_\infty.$$

Hence $\{H_{p_n}\}$ converges to H_ϕ in norm as n approaches ∞. Therefore H_ϕ is a norm limit of finite-rank Hankel operators and is thus compact.

In the other direction, assume that H_ϕ is compact. We will show that there is a continuous function ψ such that $H_\psi = H_\phi$.

Since H_ϕ is compact and $\{U^{*n}\}$ converges to 0 pointwise, it follows that $\{U^{*n}H_\phi\}$ converges to 0 in norm (see Exercise 1.19 in Chapter 1). The equations $U^*H_\phi = H_\phi U = H_{e^{i\theta}\phi}$ yield $U^{*n}H_\phi = H_{e^{in\theta}\phi}$. Thus $\{\|H_{e^{in\theta}\phi}\|\}$ converges to 0. Nehari's theorem (Theorem 4.1.13) implies that, for each natural number n, there is an h_n in $e^{i\theta}\widetilde{\boldsymbol{H}}^\infty$ such that

$$\|H_{e^{in\theta}\phi}\| = \|e^{in\theta}\phi + h_n\|_\infty.$$

Thus $\{\|e^{in\theta}\phi + h_n\|_\infty\}$ converges to 0. But

$$\|e^{in\theta}\phi + h_n\|_\infty = \|e^{in\theta}(\phi + e^{-in\theta}h_n)\|_\infty = \|\phi + e^{-in\theta}h_n\|_\infty.$$

Hence $\{\|\phi + e^{-in\theta}h_n\|_\infty\}$ converges to 0.

Observe that the coanalytic part of each function $e^{-in\theta}h_n$ has only a finite number of Fourier coefficients different from 0. Therefore the Hankel operator $H_{e^{-in\theta}h_n}$ has a matrix with only a finite number of entries different from 0. Since

$$\|H_\phi - H_{-e^{-in\theta}h_n}\| = \|H_{\phi+e^{-in\theta}h_n}\| \le \|\phi + e^{-in\theta}h_n\|_\infty,$$

the compact operator H_ϕ is the norm limit of the finite-rank Hankel operators $H_{-e^{-in\theta}h_n}$. Let α_0 be the function 0 and, for each natural number n, let α_n denote the coanalytic part of $-e^{-in\theta}h_n$.

Since $\{H_{\alpha_n}\}$ converges to H_ϕ, we can assume, by choosing a subsequence if necessary, that

$$\|H_{\alpha_{n+1}} - H_{\alpha_n}\| \le \frac{1}{2^n}.$$

Since the matrix of $H_{\alpha_{n+1}-\alpha_n}$ has only a finite number of entries different from 0, Corollary 4.2.7 implies that we can find a function g_n continuous on S^1 such that $H_{g_n} = H_{\alpha_{n+1}-\alpha_n}$ and

$$\|g_n\|_\infty = \|H_{\alpha_{n+1}-\alpha_n}\| = \|H_{\alpha_{n+1}} - H_{\alpha_n}\|.$$

Since $\|g_n\|_\infty \le \frac{1}{2^n}$ for all n, the Weierstrass M-test shows that the function ψ defined by

$$\psi(e^{i\theta}) = \sum_{k=0}^{\infty} g_k(e^{i\theta})$$

is continuous on S^1. We need only show that $H_\phi = H_\psi$, which is equivalent to establishing that the coanalytic parts of ϕ and ψ are equal.

Note that g_k and $\alpha_{k+1} - \alpha_k$ have the same Fourier coefficients for all nonpositive indices. Thus, for each nonpositive integer s,

$$
\begin{aligned}
(\psi, e^{is\theta}) &= \sum_{k=0}^{\infty} (g_k, e^{is\theta}) \\
&= \sum_{k=0}^{\infty} \left((\alpha_{k+1}, e^{is\theta}) - (\alpha_k, e^{is\theta}) \right) \\
&= \lim_{n\to\infty} (\alpha_{n+1}, e^{is\theta}) \\
&= \lim_{n\to\infty} (\alpha_n, e^{is\theta}).
\end{aligned}
$$

Recall that α_n is the coanalytic part of $-e^{-in\theta}h_n$. Since $\{-e^{-in\theta}h_n\}$ converges to ϕ in $\boldsymbol{L}^\infty$, it follows immediately that

$$\lim_{n\to\infty} (\alpha_n, e^{is\theta}) = (\phi, e^{is\theta})$$

for each nonpositive integer s. Thus the coanalytic parts of ψ and ϕ coincide, so $H_\psi = H_\phi$. □

Notation 4.3.2. We denote the set of all continuous complex-valued functions in S^1 by $\boldsymbol{C}$. The set $\widetilde{\boldsymbol{H}}^\infty + \boldsymbol{C}$ is the collection of all sums of a function in $\widetilde{\boldsymbol{H}}^\infty$ and an element of $\boldsymbol{C}$.

Hartman's theorem can be restated as follows.

Corollary 4.3.3 (Hartman's Theorem). *The Hankel operator H_ϕ is compact if and only if ϕ is in $\widetilde{\boldsymbol{H}}^\infty + \boldsymbol{C}$.*

Proof. If H_ϕ is compact, the previous version of Hartman's theorem (Theorem 4.3.1) implies that there is a continuous function ψ such that $H_\phi = H_\psi$. Then ϕ and ψ have the same coanalytic parts, so $\phi - \psi$ is a function in $\widetilde{\boldsymbol{H}}^\infty$. Thus ϕ is in $\widetilde{\boldsymbol{H}}^\infty + \boldsymbol{C}$.

Conversely, if $\phi = f + \psi$ with f in $\widetilde{\boldsymbol{H}}^\infty$ and ψ continuous, then ϕ has the same coanalytic part as the continuous function obtained from ψ by adding the constant term of f. Therefore H_ϕ is compact by the earlier version of Hartman's theorem (Theorem 4.3.1). □

4.4 Self-Adjointness and Normality of Hankel Operators

Recall that $\breve{f}$ is defined by $\breve{f}(e^{i\theta}) = f(e^{-i\theta})$ (Notation 4.2.1). We now require some additional related functions.

Notation 4.4.1. For each function f in $\boldsymbol{L}^2$, we define:

(i) the function $\overline{f}$ by

$$\overline{f}(e^{i\theta}) = \overline{f(e^{i\theta})}, \text{ and}$$

(ii) the function f^* as the function $\breve{\overline{f}}$; i.e.,

$$f^*(e^{i\theta}) = \overline{f(e^{-i\theta})}.$$

It should be noted that $f \in \widetilde{\boldsymbol{H}}^2$ if and only if $f^* \in \widetilde{\boldsymbol{H}}^2$. In fact,

$$\text{if} \quad f(e^{i\theta}) = \sum_{n=-\infty}^{\infty} a_n e^{in\theta}, \quad \text{then} \quad f^*(e^{i\theta}) = \sum_{n=-\infty}^{\infty} \overline{a_n} e^{in\theta}.$$

Also

$$\|f\| = \|\breve{f}\| = \|\overline{f}\| = \|f^*\|.$$

For $f \in \boldsymbol{L}^\infty$,

$$\|f\|_\infty = \|\breve{f}\|_\infty = \|\overline{f}\|_\infty = \|f^*\|_\infty.$$

There is a special relationship between Hankel operators and their adjoints.

Theorem 4.4.2. *If H is a Hankel operator, then $H^*f^* = (Hf)^*$ for every $f \in \widetilde{\boldsymbol{H}}^2$. In particular, $\|H^*f^*\| = \|Hf\|$ for every f in $\widetilde{\boldsymbol{H}}^2$.*

Proof. Recall that the matrix of A^* with respect to a given orthonormal basis can be obtained from that of A by taking the "conjugate transpose". Since the transpose of a Hankel matrix H is itself, the matrix of H^* is obtained from that of H by simply conjugating each of its entries. That $H^*f^* = (Hf)^*$ follows immediately from this fact. □

It is easy to describe the adjoint of a Hankel operator. If the Hankel operator H has matrix

$$\begin{pmatrix} a_0 & a_1 & a_2 & a_3 & \cdots \\ a_1 & a_2 & a_3 & ⋰ & \\ a_2 & a_3 & ⋰ & & \\ a_3 & ⋰ & & & \\ ⋰ & & & & \\ \vdots & & & & \end{pmatrix},$$

then the matrix of H^*, the conjugate transpose of the matrix of H, is simply

$$\begin{pmatrix} \overline{a_0} & \overline{a_1} & \overline{a_2} & \overline{a_3} & \cdots \\ \overline{a_1} & \overline{a_2} & \overline{a_3} & \cdot^{\cdot^{\cdot}} & \\ \overline{a_2} & \overline{a_3} & \cdot^{\cdot^{\cdot}} & & \\ \overline{a_3} & \cdot^{\cdot^{\cdot}} & & & \\ \vdots & & & & \end{pmatrix}.$$

Theorem 4.4.3. *For every ϕ in $\boldsymbol{L}^\infty$, $H_\phi^* = H_{\phi^*}$. Moreover, if H_ϕ is self-adjoint, then there exists a ψ in $\boldsymbol{L}^\infty$ such that $\psi = \psi^*$ a.e. and $H_\phi = H_\psi$.*

Proof. The fact that $H_\phi^* = H_{\phi^*}$ follows from the matrix representation of H_ϕ as indicated above. Alternatively, it is also a consequence of the following computation. For any f and g in $\boldsymbol{H}^2$,

$$(H_\phi^* f, g) = (f, H_\phi g) = (f, PJM_\phi g) = (f, JM_\phi g) = (M_{\overline{\phi}} Jf, g).$$

It is easily seen that $M_{\overline{\phi}} J = JM_{\phi^*}$. Therefore

$$(H_\phi^* f, g) = (JM_{\phi^*} f, g) = (PJM_{\phi^*} f, g) = (H_{\phi^*} f, g).$$

In particular, $H_\phi^* = H_\phi$ if and only if ϕ and ϕ^* have the same coanalytic parts. This is, of course, equivalent to the entries in the standard matrix for H_ϕ being real, which is obviously equivalent to the self-adjointness of H_ϕ.

It remains to be shown that H_ϕ self-adjoint implies that there exists a ψ in $\boldsymbol{L}^\infty$ such that $H_\phi = H_\psi$ and $\psi = \psi^*$. Since ϕ and ϕ^* have the same coanalytic parts, there is an h in $e^{i\theta}\widetilde{\boldsymbol{H}^2}$ such that $\phi - \phi^* = h$.

Note that $h^* = -h$ and $h \in \boldsymbol{L}^\infty$. Define ψ in $\boldsymbol{L}^\infty$ by $\psi = \phi - \frac{1}{2}h$. Then

$$\psi - \psi^* = \left(\phi - \frac{1}{2}h\right) - \left(\phi^* - \frac{1}{2}h^*\right) = \phi - \phi^* - h = 0,$$

and thus $\psi = \psi^*$. Clearly, $H_\psi = H_\phi$. □

We have seen that it is fairly rare that the product of two Toeplitz operators is a Toeplitz operator. It is even rarer that the product of two Hankel operators is Hankel; in fact, that occurs only if both Hankel operators are multiples of the same rank-one Hankel operator. The following lemma is useful in other contexts as well as for establishing this.

Lemma 4.4.4. *If H_ϕ and H_ψ are Hankel operators and U is the unilateral shift, then*

$$H_\phi H_\psi - U^* H_\phi H_\psi U = (P\breve{\phi}) \otimes (P\overline{\psi}),$$

where P is the projection of $\boldsymbol{L}^2$ onto $\widetilde{\boldsymbol{H}}^2$.

Proof. Note that

$$\begin{aligned} H_\phi H_\psi - U^* H_\phi H_\psi U &= H_\phi H_\psi - H_\phi U U^* H_\psi \qquad \text{(by Theorem 4.1.7)} \\ &= H_\phi (I - UU^*) H_\psi. \end{aligned}$$

Recall that $I - UU^*$ is the projection of $\widetilde{\boldsymbol{H}}^2$ onto the constant functions; i.e., $I - U^*U = e_0 \otimes e_0$. Therefore

$$\begin{aligned} H_\phi H_\psi - U^* H_\phi H_\psi U &= H_\phi (e_0 \otimes e_0) H_\psi \\ &= (H_\phi e_0) \otimes (H_{\psi^*} e_0) \\ &= (PJ\phi) \otimes (PJ\psi^*) \\ &= (P\breve{\phi}) \otimes (P\overline{\psi}). \end{aligned}$$

□

Theorem 4.4.5. *The product of two nonzero Hankel operators is a Hankel operator if and only if both of the operators are constant multiples of the same Hankel operator of rank 1.*

Proof. We have seen that every Hankel operator of rank 1 has the form $k_{\overline{w}} \otimes k_w$ for some w in $\mathbb{D}$, where k_w is the kernel function given by $k_w(z) = \frac{1}{1-\overline{w}z}$ (Theorem 4.2.2). Thus, to prove the first implication of the theorem, it suffices to show that the square of such an operator is a multiple of itself. Let f be in $\widetilde{\boldsymbol{H}}^2$. Then

$$\begin{aligned} (k_{\overline{w}} \otimes k_w)\ (k_{\overline{w}} \otimes k_w) f &= (k_{\overline{w}} \otimes k_w)\ (f, k_w) k_{\overline{w}} \\ &= (f, k_w)\ (k_{\overline{w}} \otimes k_w) k_{\overline{w}} \\ &= (f, k_w)\ (k_{\overline{w}}, k_w) k_{\overline{w}} \\ &= (k_{\overline{w}}, k_w)\ (f, k_w) k_{\overline{w}} \\ &= (k_{\overline{w}}, k_w)\ (k_{\overline{w}} \otimes k_w) f. \end{aligned}$$

For the converse, assume that H_ϕ, H_ψ, and $H_\phi H_\psi$ are all Hankel operators. Then $U^* H_\phi H_\psi U$ is also a Hankel operator (Theorem 4.1.8). By the above lemma (Lemma 4.4.4),

$$H_\phi H_\psi - U^* H_\phi H_\psi U = (P\breve{\phi}) \otimes (P\overline{\psi}).$$

Since the difference of two Hankel operators is Hankel, it follows that the operator $(P\breve{\phi}) \otimes (P\overline{\psi})$ is Hankel.

If $(P\breve{\phi}) \otimes (P\overline{\psi})$ was zero, we would have either $P\breve{\phi} = 0$ or $P\overline{\psi} = 0$. It would then follow that H_ϕ or H_ψ is zero, which they are not. Thus the operator $(P\breve{\phi}) \otimes (P\overline{\psi})$ is a Hankel operator of rank 1.

By Theorem 4.2.2, then, we have

$$(P\breve{\phi}) \otimes (P\overline{\psi}) = ck_{\overline{w}} \otimes k_w$$

for some $w \in \mathbb{D}$ and a nonzero constant c. Therefore there exist constants a and b such that

$$P\breve{\phi} = a\, k_{\overline{w}} \quad \text{and} \quad P\overline{\psi} = b\, k_w,$$

by Theorem 1.2.28. Note that $P(\overline{\psi}) = bk_w$ implies that $P(\overline{\psi}^*) = \overline{b}k_w^*$. However, $\overline{\psi}^* = \breve{\psi}$ and $k_w^* = k_{\overline{w}}$. Therefore we have

$$P\breve{\phi} = a\, k_{\overline{w}} \quad \text{and} \quad P\breve{\psi} = \overline{b}\, k_{\overline{w}}.$$

Hence the coanalytic parts of ϕ and ψ are both multiples of $\breve{k}_{\overline{w}}$. Thus, by Theorem 4.2.2, both of H_ϕ and H_ψ are multiples of the rank-one operator $k_{\overline{w}} \otimes k_w$. □

Corollary 4.4.6. *The product of two Hankel operators is a Toeplitz operator only if at least one of the Hankel operators is* 0.

Proof. By Lemma 4.4.4,

$$H_\phi H_\psi - U^* H_\phi H_\psi U = (P\breve{\phi}) \otimes (P\overline{\psi}).$$

If $H_\phi H_\psi$ is a Toeplitz operator, then $U^* H_\phi H_\psi U = H_\phi H_\psi$ (by Corollary 3.2.7), so

$$(P\breve{\phi}) \otimes (P\overline{\psi}) = 0.$$

Therefore at least one of $(P\breve{\phi})$ and $(P\overline{\psi})$ is 0. As in the previous theorem, it follows that either H_ϕ or H_ψ is 0. □

Corollary 4.4.7. *The product of two Hankel operators is* 0 *if and only if one of them is* 0.

Proof. If the product of two Hankel operators is the Toeplitz operator 0, the previous corollary implies that at least one of the Hankel operators is zero. □

The question arises of when two Hankel operators commute with each other. It turns out that this happens only if one is a multiple of the other.

Theorem 4.4.8. *Let ϕ and ψ be in $\mathbf{L}^\infty$ and suppose that $H_\psi \neq 0$. If H_ϕ and H_ψ commute, then there exists a complex number c such that $H_\phi = cH_\psi$.*

Proof. When $H_\phi \neq 0$, by Lemma 4.4.4,

$$H_\phi H_\psi - U^* H_\phi H_\psi U = (P\breve{\phi}) \otimes (P\overline{\psi}),$$

and

$$H_\psi H_\phi - U^* H_\psi H_\phi U = (P\breve{\psi}) \otimes (P\overline{\phi}).$$

Therefore $H_\psi H_\phi = H_\phi H_\psi$ implies that

$$(P\breve{\phi}) \otimes (P\overline{\psi}) = (P\breve{\psi}) \otimes (P\overline{\phi}).$$

In the case that $H_\phi \neq 0$, all the vectors defining this rank-one operator are different from 0. It follows that there exists a complex number c different from 0 such that

$$P\breve{\phi} = cP(\breve{\psi}).$$

Therefore $H_\phi = cH_\psi$. □

There are very few normal Hankel operators.

Corollary 4.4.9. *Every normal Hankel operator is a multiple of a self-adjoint Hankel operator.*

Proof. Let H be a normal Hankel operator; i.e., $HH^* = H^*H$. If $H = 0$, the result is trivial. In the other case, by the previous theorem, there is a constant c such that $H = cH^*$. Since $\|H\| = \|H^*\|$, we have $|c| = 1$.

Let $c = e^{i\theta}$; it follows that

$$(e^{-i\theta/2}H)^* = (e^{-i\theta/2}H),$$

so $e^{-i\theta/2}H$ is self-adjoint and the result follows. □

There is a generalization of normality that is sometimes studied.

Definition 4.4.10. The bounded operator A is *hyponormal* if $\|Af\| \geq \|A^*f\|$ for every vector $f \in \mathcal{H}$.

There are also very few hyponormal Hankel operators.

Theorem 4.4.11. *Every hyponormal Hankel operator is normal.*

Proof. By Theorem 4.4.2, it follows that $\|H^*f^*\| = \|Hf\|$ for every $f \in \widetilde{\boldsymbol{H}}^2$, where f^* is the vector whose coefficients are the conjugates of those of f (Notation 4.4.1). Applying this to f^* yields $\|H^*f\| = \|Hf^*\|$.

If H is hyponormal, then $\|Hf\| \geq \|H^*f\|$ for every f. Therefore, $\|Hf^*\| \geq \|H^*f^*\|$. By the above equations, this yields $\|H^*f\| \geq \|Hf\|$. But $\|Hf\| \geq \|H^*f\|$, since H is hyponormal. Hence $\|H^*f\| = \|Hf\|$ for all f, and H is normal. □

4.5 Relations Between Hankel and Toeplitz Operators

There are some interesting relations between the Hankel and Toeplitz operators with symbols ϕ, ψ, and $\phi\psi$. One consequence of these formulas is a precise determination of when a Hankel and a Toeplitz operator commute with each other.

Theorem 4.5.1. *Let ϕ and ψ be in $\boldsymbol{L}^\infty$. Then*

$$H_{e^{i\theta}\breve{\phi}} H_{e^{i\theta}\psi} = T_{\phi\psi} - T_\phi T_\psi.$$

Proof. The flip operator, J, and the projection onto $\widetilde{\boldsymbol{H}}^2$, P, satisfy the following equation:

$$JPJ = M_{e^{i\theta}}(I - P)M_{e^{-i\theta}}.$$

(This can easily be verified by applying each side to the basis vectors $\{e^{in\theta}\}$.) Thus

$$\begin{aligned}
H_{e^{i\theta}\breve{\phi}} H_{e^{i\theta}\psi} &= (PJM_{e^{i\theta}\breve{\phi}})\,(PJM_{e^{i\theta}\psi}) \\
&= P\,(M_{e^{-i\theta}\phi}J)\,(PJM_{e^{i\theta}\psi}) \qquad \text{since } JM_{e^{i\theta}\breve{\phi}} = M_{e^{-i\theta}\phi}J \\
&= PM_\phi M_{e^{-i\theta}}\,(JPJ)\,M_{e^{i\theta}}M_\psi \\
&= PM_\phi M_{e^{-i\theta}}\,(M_{e^{i\theta}}(I-P)M_{e^{-i\theta}})\,M_{e^{i\theta}}M_\psi \\
&= PM_\phi(I-P)M_\psi \\
&= (PM_\phi M_\psi) - (PM_\phi)\,(PM_\psi) \\
&= T_{\phi\psi} - T_\phi T_\psi.
\end{aligned}$$

□

It is easy to rephrase the above theorem to express the product of any two Hankel operators in terms of Toeplitz operators.

Corollary 4.5.2. *If ϕ and ψ are in $\boldsymbol{L}^\infty$, then*

$$H_\phi H_\psi = T_{\breve{\phi}\psi} - T_{e^{i\theta}\breve{\phi}} T_{e^{-i\theta}\psi}.$$

Proof. By the previous theorem,

$$H_{e^{i\theta}\breve{\alpha}} H_{e^{i\theta}\beta} = T_{\alpha\beta} - T_\alpha T_\beta$$

for α and β in $\boldsymbol{L}^\infty$. Let $\alpha = e^{i\theta}\breve{\phi}$ and $\beta = e^{-i\theta}\psi$. Making this substitution in the equation above gives the result. □

One consequence of this corollary is another proof of the following (cf. Corollary 4.4.6).

Corollary 4.5.3. *If the product of two Hankel operators is Toeplitz, then at least one of the Hankel operators is* 0.

Proof. If $H_\phi H_\psi$ is a Toeplitz operator, then since the sum of two Toeplitz operators is Toeplitz, it follows from the previous corollary that $T_{e^{i\theta}\breve{\phi}} T_{e^{-i\theta}\psi}$ is a Toeplitz operator. Thus either $e^{-i\theta}\psi$ is analytic or $e^{i\theta}\breve{\phi}$ is coanalytic (Theorem 3.2.11), so $H_\psi = 0$ or $H_\phi = 0$. □

It should also be noted that Theorem 4.5.1 shows that the following facts, which we previously obtained independently, are equivalent to each other:

- If the product of two Hankel operators is zero, then one of them is zero (Corollary 4.4.7).
- If $T_{\phi\psi} = T_\phi T_\psi$, then either ϕ is coanalytic or ψ is analytic (Theorem 3.2.11).

Another important equation relating Hankel and Toeplitz operators is the following.

Theorem 4.5.4. *Let ϕ and ψ be in $\boldsymbol{L}^\infty$. Then*

$$T_{\breve{\phi}} H_{e^{i\theta}\psi} + H_{e^{i\theta}\phi} T_\psi = H_{e^{i\theta}\phi\psi}.$$

Proof. This follows from a computation similar to that in the proof of Theorem 4.5.1. Using $JM_\phi J = M_{\breve{\phi}}$ and $JPJ = M_{e^{i\theta}}(I-P)M_{e^{-i\theta}}$, we get

$$\begin{aligned}
T_{\breve{\phi}} H_{e^{i\theta}\psi} &= (PM_{\breve{\phi}})(PJM_{e^{i\theta}\psi}) \\
&= P(JM_{\phi}J)(PJM_{e^{i\theta}}M_{\psi}) \\
&= PJM_{\phi}(JPJ)\,M_{e^{i\theta}}M_{\psi} \\
&= PJM_{\phi}(M_{e^{i\theta}}(I-P)M_{e^{-i\theta}})\,M_{e^{i\theta}}M_{\psi} \\
&= PJM_{e^{i\theta}\phi}(I-P)M_{\psi} \\
&= (PJM_{e^{i\theta}\phi}M_{\psi}) - (PJM_{e^{i\theta}\phi})(PM_{\psi}) \\
&= H_{e^{i\theta}\phi\psi} - H_{e^{i\theta}\phi}T_{\psi}.
\end{aligned}$$

□

Under certain circumstances, the product of a Hankel operator and a Toeplitz operator is a Hankel operator.

Corollary 4.5.5. *(i) If ψ is in $\widetilde{\boldsymbol{H}}^{\infty}$, then $H_{\phi}T_{\psi} = H_{\phi\psi}$.*

(ii) If ψ is in $\widetilde{\boldsymbol{H}}^{\infty}$, then $T_{\breve{\psi}}H_{\phi} = H_{\psi\phi}$.

Proof. Recall from the previous theorem that, for α and β in $\boldsymbol{L}^{\infty}$,

$$T_{\breve{\alpha}}H_{e^{i\theta}\beta} + H_{e^{i\theta}\alpha}T_{\beta} = H_{e^{i\theta}\alpha\beta}.$$

Taking $\alpha = e^{-i\theta}\phi$ and $\beta = \psi$ gives *(i)*, since $H_{e^{i\theta}\beta} = 0$. Taking $\alpha = \psi$ and $\beta = e^{-i\theta}\phi$ we obtain *(ii)*, since $H_{e^{i\theta}\alpha} = 0$. □

We have seen that Toeplitz operators rarely commute with each other (Theorem 3.2.13) and that Hankel operators rarely commute with each other (Theorem 4.4.8). We now consider the question of determining when a Hankel operator commutes with a Toeplitz operator. This is also quite rare.

Theorem 4.5.6. *Suppose neither of the Toeplitz operator T_{ϕ} and the Hankel operator H is a multiple of the identity. Then $HT_{\phi} = T_{\phi}H$ if and only if H is a multiple of $H_{e^{i\theta}\phi}$ and both of the functions $\phi + \breve{\phi}$ and $\phi\breve{\phi}$ are constant functions.*

Proof. First suppose that $\phi + \breve{\phi} = c$ and $\phi\breve{\phi} = d$ for complex numbers c and d. Theorem 4.5.4 states that

$$T_{\phi}H_{e^{i\theta}\phi} + H_{e^{i\theta}\breve{\phi}}T_{\phi} = H_{e^{i\theta}\phi\breve{\phi}}.$$

Since $\phi\breve{\phi} = d$, it follows that

$$T_\phi H_{e^{i\theta}\phi} + H_{e^{i\theta}\breve{\phi}} T_\phi = H_{e^{i\theta}\phi\breve{\phi}} = H_{de^{i\theta}}.$$

Since $de^{i\theta}$ is in $e^{i\theta}\widetilde{\boldsymbol{H}}^2$, it follows that $H_{de^{i\theta}} = 0$, and thus that

$$T_\phi H_{e^{i\theta}\phi} + H_{e^{i\theta}\breve{\phi}} T_\phi = 0.$$

Also, since $\phi + \breve{\phi} = c$, we have $H_{e^{i\theta}\breve{\phi}} = H_{ce^{i\theta} - e^{i\theta}\phi}$, and therefore

$$T_\phi H_{e^{i\theta}\phi} + H_{ce^{i\theta}} T_\phi - H_{e^{i\theta}\phi} T_\phi = 0.$$

Since $ce^{i\theta}$ is in $e^{i\theta}\widetilde{\boldsymbol{H}}^2$, $H_{ce^{i\theta}} = 0$. Therefore

$$T_\phi H_{e^{i\theta}\phi} = H_{e^{i\theta}\phi} T_\phi.$$

It follows that if H is a multiple of $H_{e^{i\theta}\phi}$, then H commutes with T_ϕ.

To prove the converse, suppose that $HT_\phi = T_\phi H$. Multiplying $T_\phi H$ on the right by U, using the fact that $U^*H = HU$, noticing that $H_{e_0} = PJM_{e_0} = PJ = e_0 \otimes e_0$, and using Theorem 4.5.1, we get

$$\begin{aligned}
T_\phi HU &= T_\phi U^* H \\
&= T_\phi T_{e^{-i\theta}} H \\
&= \left(T_{\phi e^{-i\theta}} - H_{e^{i\theta}\breve{\phi}} H_{e^{i\theta}e^{-i\theta}}\right) H \\
&= \left(T_{e^{-i\theta}} T_\phi - H_{e^{i\theta}\breve{\phi}} H_{e_0}\right) H \\
&= U^* T_\phi H - H_{e^{i\theta}\breve{\phi}} (e_0 \otimes e_0) H \\
&= U^* T_\phi H - \left(H_{e^{i\theta}\breve{\phi}} e_0\right) \otimes (H^* e_0).
\end{aligned}$$

Performing similar computations beginning with HT_ϕ yields

$$\begin{aligned}
HT_\phi U &= HT_\phi T_{e^{i\theta}} \\
&= HT_{e^{i\theta}\phi} \\
&= H\left(T_{e^{i\theta}} T_\phi + H_{e^{i\theta}e^{-i\theta}} H_{e^{i\theta}\phi}\right) \\
&= H\left(UT_\phi + H_{e_0} H_{e^{i\theta}\phi}\right) \\
&= HUT_\phi + H(e_0 \otimes e_0) H_{e^{i\theta}\phi} \\
&= U^* HT_\phi + (He_0) \otimes \left(H_{e^{i\theta}\phi^*} e_0\right).
\end{aligned}$$

Since $HT_\phi = T_\phi H$, it follows that

$$-\left(H_{e^{i\theta}\breve{\phi}} e_0\right) \otimes (H^* e_0) = (He_0) \otimes \left(H_{e^{i\theta}\phi^*} e_0\right).$$

Since H is not zero, neither He_0 nor H^*e_0 is zero.

If either one of $H_{e^{i\theta}\breve{\phi}}e_0$ and $H_{e^{i\theta}\phi^*}e_0$ were 0, then the other would also be 0. It would then follow that $\breve{\phi}$ and ϕ^* are both in $\widetilde{\boldsymbol{H}}^2$, which happens only if ϕ is a constant. The assumption that T_ϕ is not a multiple of the identity implies that ϕ is not a constant function. It follows that each of the four vectors occurring in the equation

$$-\left(H_{e^{i\theta}\breve{\phi}}e_0\right) \otimes (H^*e_0) = (He_0) \otimes \left(H_{e^{i\theta}\phi^*}e_0\right)$$

is different from 0.

Thus Theorem 1.2.28 implies that there is a complex number $c \neq 0$ such that

$$-cH_{e^{i\theta}\breve{\phi}}\, e_0 = He_0 \qquad \text{and} \qquad H^*e_0 = \bar{c}H_{e^{i\theta}\phi^*}\, e_0.$$

Therefore $H = -cH_{e^{i\theta}\breve{\phi}}$ and $H^* = \bar{c}H_{e^{i\theta}\phi^*}$. Taking the adjoint of both sides of the latter equation gives $H = cH_{e^{i\theta}\phi}$, which proves the first conclusion of the theorem.

It follows from the above that

$$cH_{e^{i\theta}\phi} = -cH_{e^{i\theta}\breve{\phi}}.$$

Therefore

$$H_{e^{i\theta}\phi} + H_{e^{i\theta}\breve{\phi}} = 0.$$

But

$$H_{e^{i\theta}\phi} + H_{e^{i\theta}\breve{\phi}} = H_{e^{i\theta}(\phi+\breve{\phi})},$$

which implies that $\phi + \breve{\phi}$ is in $\widetilde{\boldsymbol{H}}^2$. Let $\psi = \phi + \breve{\phi}$. Then $\psi = \breve{\psi}$ and ψ is in $\widetilde{\boldsymbol{H}}^2$, which implies that ψ is a constant, which proves the second conclusion of the theorem.

We must show that $\phi\breve{\phi}$ is a constant. Notice that the fact that $\phi + \breve{\phi}$ is constant implies that $T_{\breve{\phi}}$ is a translate of T_ϕ, so it follows that H also commutes with $T_{\breve{\phi}}$. Applying Theorem 4.5.1 gives

$$H^2_{e^{i\theta}\phi} = T_{\breve{\phi}\phi} - T_{\breve{\phi}}T_\phi.$$

Since H is a multiple of $H_{e^{i\theta}\phi}$, it commutes with the left-hand side of the above equation. Since it also commutes with T_ϕ and $T_{\breve{\phi}}$, it follows that H commutes with $T_{\breve{\phi}\phi}$. By the above proof, $\breve{\phi}\phi + (\breve{\phi}\phi)\breve{}$ is constant. But $(\phi\breve{\phi})\breve{} = \phi\breve{\phi}$, so $\phi\breve{\phi}$ is constant. □

This theorem suggests the question of determining the functions ϕ in $\boldsymbol{L}^\infty$ such that both $\phi + \breve{\phi}$ and $\phi\breve{\phi}$ are constants.

Theorem 4.5.7. *For E a measurable subset of S^1, let E^* denote the set of complex conjugates of elements of E, and let E^c denote the complement of E in S^1. For $\phi \in \mathbf{L}^2$, $\phi + \breve{\phi}$ and $\phi\breve{\phi}$ are both constant functions if and only if there are complex numbers a and b such that $\phi(e^{i\theta}) = a\chi_E(e^{i\theta}) + b$, where χ_E is the characteristic function of a measurable set of S^1 such that $m(E^* \setminus E^c) + m(E^c \setminus E^*) = 0$.*

Proof. If $\phi(e^{i\theta}) = a\chi_E(e^{i\theta}) + b$ as in the statement of the theorem, then $\breve{\phi}(z) = a\chi_{E^*}(e^{i\theta}) + b$ and

$$\begin{aligned}
\phi(e^{i\theta}) + \breve{\phi}(e^{i\theta}) &= a(\chi_E(e^{i\theta}) + \chi_{E^*}(e^{i\theta})) + 2b \\
&= a(\chi_{E\cup E^*}(e^{i\theta}) - \chi_{E\cap E^*}(e^{i\theta})) + 2b \\
&= a(\chi_{S^1}(e^{i\theta}) - \chi_{\emptyset}(e^{i\theta})) + 2b \\
&= a(1 + 0) + 2b \\
&= a + 2b,
\end{aligned}$$

because E^* and E^c coincide except possibly on a set of measure 0. Similarly,

$$\begin{aligned}
\phi(e^{i\theta})\breve{\phi}(e^{i\theta}) &= (a\chi_E(e^{i\theta}) + b)(a\chi_{E^*}(e^{i\theta}) + b) \\
&= a^2\chi_E(e^{i\theta})\chi_{E^*}(e^{i\theta}) + ab(\chi_E(e^{i\theta}) + \chi_{E^*}(e^{i\theta})) + b^2 \\
&= a^2\chi_{E\cap E^*}(e^{i\theta}) + ab + b^2 \\
&= a^2 0 + ab + b^2,
\end{aligned}$$

so both functions are constant.

Conversely, suppose that $\phi + \breve{\phi}$ and $\phi\breve{\phi}$ are constant. We first establish a special case. Suppose that $\phi + \breve{\phi} = 1$ and $\phi\breve{\phi} = 0$. Then

$$\phi = \phi(\phi + \breve{\phi}) = \phi^2 + \phi\breve{\phi} = \phi^2,$$

which implies that $\phi = \chi_E$ for some measurable subset $E \subset S^1$. Since

$$0 = \phi(e^{i\theta})\breve{\phi}(e^{i\theta}) = \chi_E(e^{i\theta})\chi_{E^*}(e^{i\theta}) = \chi_{E\cap E^*}(e^{i\theta}),$$

it follows that $m(E \cap E^*) = 0$. Analogously,

$$\begin{aligned}
1 = \phi(e^{i\theta}) + \breve{\phi}(e^{i\theta}) &= \chi_E(e^{i\theta}) + \chi_{E^*}(e^{i\theta}) \\
&= \chi_{E\cup E^*}(e^{i\theta}) - \chi_{E\cap E^*}(e^{i\theta}) = \chi_{E\cup E^*}(e^{i\theta}),
\end{aligned}$$

so $m(E \cup E^*) = 1$. Thus $m(E^* \setminus E^c) + m(E^c \setminus E^*) = 0$ and the conclusion follows with $a = 1$ and $b = 0$.

To deal with the general situation, suppose $\phi + \breve{\phi} = c$ and $\phi\breve{\phi} = d$. Let b be a solution of the equation $x^2 - cx + d = 0$ and let $a = c - 2b$.

If $a \neq 0$, then define $\psi(e^{i\theta}) = \frac{\phi(e^{i\theta})-b}{a}$. Then

$$\psi(e^{i\theta}) + \breve{\psi}(e^{i\theta}) = \frac{\phi(e^{i\theta}) - b}{a} + \frac{\breve{\phi}(e^{i\theta}) - b}{a} = \frac{c - 2b}{a} = \frac{a}{a} = 1$$

and

$$\psi(e^{i\theta})\breve{\psi}(e^{i\theta}) = \left(\frac{\phi(e^{i\theta}) - b}{a}\right)\left(\frac{\breve{\phi}(e^{i\theta}) - b}{a}\right) = \frac{d - bc + b^2}{a^2} = \frac{0}{a^2} = 0.$$

Therefore, by the previous case, $\psi = \chi_E$, where E is as described above. Thus $\phi(e^{i\theta}) = a\psi(e^{i\theta}) + b = a\chi_E(e^{i\theta}) + b$.

If $a = 0$, then let $\psi(e^{i\theta}) = \phi(e^{i\theta}) - b$. Then

$$\psi(e^{i\theta}) + \breve{\psi}(e^{i\theta}) = (\phi(e^{i\theta}) - b) + (\breve{\phi}(e^{i\theta}) - b) = c - 2b = 0$$

and

$$\psi(e^{i\theta})\breve{\psi}(e^{i\theta}) = (\phi(e^{i\theta}) - b)(\breve{\phi}(e^{i\theta}) - b) = d - bc + b^2 = 0.$$

Thus $\breve{\psi} = -\psi$ and $-\psi^2 = 0$, which implies that $\psi = 0$; i.e., $\phi(z) = b$, as desired. □

4.6 Exercises

4.1. Show that $\|f\| = \|\breve{f}\| = \|\overline{f}\| = \|f^*\|$ for every f in $\boldsymbol{L}^2$.

4.2. Prove that, for f in $\boldsymbol{L}^\infty$, each of the functions $\breve{f}$, $\overline{f}$, and f^* is in $\boldsymbol{L}^\infty$ and

$$\|f\|_\infty = \|\breve{f}\|_\infty = \|\overline{f}\|_\infty = \|f^*\|_\infty.$$

4.3. Show that the flip operator J on $\boldsymbol{L}^2$ is unitary and self-adjoint.

4.4. For ϕ in $\boldsymbol{L}^\infty$, prove that the norm of H_ϕ is the distance from ϕ to $z\widetilde{\boldsymbol{H}^\infty}$ (i.e., show that $\|H_\phi\| = \inf\{\|\phi - zf\|_\infty \ : \ f \in \widetilde{\boldsymbol{H}^\infty}\}$).

4.5. Prove that $T_{\phi\psi} - T_\phi T_\psi$ is a compact operator if at least one of ϕ and ψ is in $\widetilde{\boldsymbol{H}^\infty} + \boldsymbol{C}$. (Hint: Use Theorem 4.5.1.)

4.6. Prove that $\widetilde{\boldsymbol{H}^\infty} + \boldsymbol{C}$ is a closed subset of $\boldsymbol{L}^\infty$. (Hint: Hartman's theorem can be used.)

4.7. Show that $\boldsymbol{H}^2$ does not contain any rational functions with poles on S^1. (Hint: It suffices to consider rational functions all of whose poles are on S^1.) Note that this establishes that rational functions that are in $\boldsymbol{H}^2$ are also in $\boldsymbol{H}^\infty$.

4.8. Prove that $JPJ = W(I - P)W^*$, where J is the flip operator, W is the bilateral shift, and P is the projection of $\boldsymbol{L}^2$ onto $\widetilde{\boldsymbol{H}^2}$.

4.9. Let J be the flip operator on $\boldsymbol{L}^2$. Show that $M_\psi J = JM_{\breve{\psi}}$ for all ψ in $\boldsymbol{L}^\infty$.

4.10. Show that the operator $k_{\overline{w}} \otimes k_w$ is Hankel by verifying that

$$U^* \left(k_{\overline{w}} \otimes k_w\right) = \left(k_{\overline{w}} \otimes k_w\right) U.$$

(Compare with Theorem 4.2.2.)

4.11. Show that, for each λ in $\mathbb{D}$, the operator

$$H = \frac{z}{(1-\lambda z)^2} \otimes \frac{1}{1-\overline{\lambda} z} + \frac{1}{1-\lambda z} \otimes \frac{z}{(1-\overline{\lambda} z)^2}$$

is a Hankel operator of rank 2. Prove that H cannot be written as the sum of two Hankel operators of rank 1.

4.12. Let $\phi(z) = \overline{z}$. Prove that the rank-two Hankel operator H_ϕ cannot be written as a finite sum of rank-one Hankel operators.

4.13. Prove that no Hankel operator is invertible. (Hint: An easy proof follows from $U^*H = HU$.)

4.14. Prove that a Hankel operator is injective if and only if it has dense range.

4.15. Prove that no Hankel operator is bounded below.

4.16. Show that the kernel of every Hankel operator H is invariant under the unilateral shift U.

4.17. Use the preceding problem to show that every Hankel operator that has a nontrivial kernel has the form $H_{z\overline{\phi}h}$, where ϕ is an inner function and h is in $\boldsymbol{H}^\infty$.

4.18. Suppose that the Hankel operator H is not injective. Prove that there exists an inner function ϕ such that the kernel of H is $\phi\widetilde{\boldsymbol{H}}^2$ and the closure of the range of H is $(\phi^*\widetilde{\boldsymbol{H}}^2)^\perp$.

4.19. An operator T is said to be a *partial isometry* if $\|Tf\| = \|f\|$ for all f in $(\operatorname{Ker} T)^\perp$. Let ϕ be an inner function. Prove that the Hankel operator $H_{z\overline{\phi}}$ is a partial isometry.

4.20. (Kronecker's Theorem) Prove that the bounded Hankel operator H has finite rank if and only if there is a finite Blaschke product B and a function h in $\widetilde{\boldsymbol{H}}^\infty$ such that $H = H_{z\overline{B}h}$.

4.21. The Hankel matrix

$$H = \left(\frac{1}{n+m+1}\right)_{n,m=0}^{\infty}$$

is called the *Hilbert matrix*. Prove that H is the matrix of the bounded Hankel operator with symbol defined by $\phi(e^{i\theta}) = i(\theta - \pi)e^{i\theta}$, for $\theta \in [0, 2\pi]$.

4.22. Prove *Hilbert's inequality*: if $\{a_n\}$ and $\{b_n\}$ are two sequences in ℓ^2, then

$$\left|\sum_{n=0}^{\infty}\sum_{m=0}^{\infty}\frac{a_n\overline{b_m}}{n+m+1}\right| \leq \pi\left(\sum_{n=0}^{\infty}|a_n|^2\right)^{1/2}\left(\sum_{m=0}^{\infty}|b_m|^2\right)^{1/2}.$$

(Hint: Use the previous problem.)

4.23. For $\lambda \in \mathbb{D}$, let k_λ be the reproducing kernel for λ defined by $k_\lambda(z) = (1 - \overline{\lambda} z)^{-1}$. Find an explicit expression for a symbol ϕ such that $H_\phi = H_{\breve{k}_\lambda}$ and $\|H_\phi\| = \|\phi\|_\infty$.

4.24. An operator T is said to be a *Hilbert–Schmidt operator* if, for some orthonormal basis $\{e_n\}$, the sum $\sum_{n=0}^{\infty} \|Te_n\|^2$ is finite. Suppose H is a bounded Hankel operator. Prove that H is Hilbert–Schmidt if and only if

$$\int_{\mathbb{D}} |\phi'(z)|^2 \, dx \, dy < \infty,$$

where ϕ is the function in $\boldsymbol{H}^2$ defined by $\widetilde{\phi} = H1$.

4.7 Notes and Remarks

Hankel matrices and operators are named after the nineteenth century mathematician Hermann Hankel, whose doctoral dissertation [97] included the study of determinants of finite Hankel matrices.

The first result concerning infinite Hankel matrices is Kronecker's theorem (Theorem 4.2.3); it was established in [110]. The alternative version of Kronecker's theorem presented in Exercise 4.20 is taken from Power [40]. Exercise 4.12 comes from Partington [36].

The earliest theorem concerning boundedness of a Hankel operator is Hilbert's inequality (Exercise 4.22 above), which is equivalent to the boundedness of the so-called Hilbert matrix (Exercise 4.21). Properties of the Hilbert matrix are elegantly discussed in Choi [70]. See [25] for another proof of Hilbert's inequality.

Nehari's theorem (Theorem 4.1.13) was originally proven in 1957 (see [120]). The proof we have given is modeled on that of [35], which also contains a third proof of the theorem based on a deep result concerning the factorization of $\boldsymbol{H}^1$ functions. The key lemma in the proof that we have presented is Parrott's theorem (Theorem 4.1.12) [125]; results related to Parrott's theorem can be found in Davis [81], Davis–Kahan–Weinberger [82], and Adamjan–Arov–Kreĭn [58].

Douglas's theorem (Theorem 4.1.9), which is used in the proof of Parrott's theorem, appeared in [85]. The proof of Parrott's theorem also depends on the Julia–Halmos theorem (Theorem 4.1.11): the fact that every operator A such that $\|A\| \leq 1$ can be so "dilated" to an isometry was proven by Julia [107], after which Halmos [95] improved the theorem to its present form. The theorem was substantially strengthened by Sz.-Nagy [160], who showed that, whenever $\|A\| \leq 1$, there exists a unitary operator U on a larger space such that the compression of U^n is A^n for every natural number n. Sz.-Nagy's theorem is the foundation of the theory of unitary dilations; see Sz.-Nagy–Foias [54].

Hartman's theorem (Theorem 4.3.1), determining when a Hankel operator is compact, was obtained in [99]. Our exposition is based on Hartman's. There are alternative proofs. In particular, if the result of Exercise 4.6 is obtained independently (as was done by Sarason [149]), Hartman's theorem is an easy consequence. Bonsall and Power [66] found a different proof of Hartman's theorem.

Theorem 4.4.5, Corollary 4.4.7, Theorem 4.4.8, Theorem 4.4.9, and Theorem 4.4.11 were pointed out by Power [141]. Exercise 4.19 is also due to Power [40]. Theorems 4.5.1 and 4.5.4 are folklore. One can also prove those theorems by regarding Hankel operators as operators from $\widetilde{\boldsymbol{H}}^2$ into $\boldsymbol{L}^2 \ominus \widetilde{\boldsymbol{H}}^2$ and considering them as compressions of matrices of multiplication operators on $\boldsymbol{L}^2$ (see, for example, [62]).

Theorems 4.5.6 and 4.5.7 were observed in [113]. The proof of Theorem 4.5.6 presented is a simplification of the one in [113], incorporating some ideas of D. Zheng and L. Robert-González (personal communications).

There are many other results in the theory of Hankel operators. Spectral properties of Hankel operators are much harder to obtain than those of Toeplitz and composition operators. A few easy results are included in the exercises above (Exercises 4.14 is a result that can be found in [40]). Some interesting theorems concerning spectra were found by Power ([136], [137], [138], [139], [140]); Power's book [40] includes a full treatment of these results as well as much additional material.

Although it is not easy to prove, it has been shown by Martínez-Avendaño and Treil [116] that every compact subset of the plane that contains zero is the spectrum of a Hankel operator. An outline of a different approach to part of the proof of this result is described in [35].

The problem of completely classifying self-adjoint Hankel operators up to unitary equivalence has been solved by Megretskiĭ, Peller, and Treil in [118]. In fact, Treil ([163], [164], [165]) and Treil and Vasyunin [166] completely classify all the moduli of Hankel operators (i.e., operators of the form $(H^*H)^{1/2}$ for H a Hankel operator).

A natural question is the determination of when a Hankel operator is in a Schatten p-class; Peller [126] completely answered this question. A fuller discussion can be found in Peller's book [38], which also contains numerous other interesting results about Hankel operators.

Applications of Hankel operators to problems in engineering are discussed in Francis [21], Helton–Merino [30], and Partington [36].

Chapter 5

Composition Operators

We introduce the class of composition operators, which are operators induced by analytic functions mapping the unit disk into itself. We discuss some of the main properties of such operators, including some aspects of their spectra.

While in the previous chapters we usually thought of $\boldsymbol{H}^2$ functions in terms of their boundary values (i.e., as functions in $\widetilde{\boldsymbol{H}}^2$, as a subset of $\boldsymbol{L}^2$), in the present chapter we primarily consider them as analytic functions on $\mathbb{D}$.

5.1 Fundamental Properties of Composition Operators

Definition 5.1.1. For each analytic function ϕ mapping the open unit disk into itself, we define the *composition operator* C_ϕ by

$$(C_\phi f)(z) = f(\phi(z))$$

for all $f \in \boldsymbol{H}^2$.

There are several points to make regarding this definition. Even though clearly $f \circ \phi$ is analytic on $\mathbb{D}$, it is not at all obvious that $f \circ \phi$ is an element of $\boldsymbol{H}^2$ for every $f \in \boldsymbol{H}^2$. It is trivially verified that C_ϕ is a linear operator. Indeed, clearly

$$\begin{aligned}(C_\phi(af + bg))(z) &= (af + bg)(\phi(z)) \\ &= af(\phi(z)) + bg(\phi(z)) \\ &= a(C_\phi f)(z) + b(C_\phi g)(z) \\ &= (aC_\phi f + bC_\phi g)(z)\end{aligned}$$

for $a, b \in \mathbb{C}$ and $f, g \in \boldsymbol{H}^2$, and thus $C_\phi(af + bg) = aC_\phi f + bC_\phi g$.

The product of two composition operators is a composition operator.

Theorem 5.1.2. *If C_ϕ and C_ψ are composition operators then $C_\phi C_\psi = C_{\psi\circ\phi}$.*

Proof. Note that

$$(C_\phi C_\psi f)(z) = (C_\phi(f\circ\psi))(z) = (f\circ\psi\circ\phi)(z) = (C_{\psi\circ\phi}f)(z),$$

and thus

$$C_\phi C_\psi = C_{\psi\circ\phi}.$$

□

Example 5.1.3. *Let $\phi(z) = z^2$. If f is in $\boldsymbol{H}^2$, then $C_\phi f$ is in $\boldsymbol{H}^2$ and, in fact, $\|C_\phi f\| = \|f\|$. Therefore C_ϕ is an isometry mapping $\boldsymbol{H}^2$ into itself.*

Proof. If f has power series $f(z) = \sum_{n=0}^\infty a_n z^n$, then $C_\phi f$ has power series $(C_\phi f)(z) = \sum_{n=0}^\infty a_n z^{2n}$. Thus $C_\phi f \in \boldsymbol{H}^2$ and all the above assertions are immediate. □

The previous example is very easy. However, it is not so easy to prove that C_ϕ maps $\boldsymbol{H}^2$ into itself if, for example, $\phi(z) = \frac{1+z}{2}$. Writing the power series decomposition is not very helpful in this case and in the case of most functions ϕ. Fortunately, we can use properties of harmonic functions to obtain a proof that every C_ϕ is a bounded linear operator mapping $\boldsymbol{H}^2$ into itself. First we need a lemma.

Lemma 5.1.4. *If $f \in \boldsymbol{H}^2$, then, for $re^{it} \in \mathbb{D}$, we have*

$$|f(re^{it})|^2 \le \frac{1}{2\pi}\int_0^{2\pi} P_r(\theta - t)|f(e^{i\theta})|^2\,d\theta.$$

Proof. Recall that, by the Poisson integral formula (Theorem 1.1.21), we have

$$f(re^{it}) = \frac{1}{2\pi}\int_0^{2\pi} P_r(\theta - t)f(e^{i\theta})\,d\theta.$$

If we define the measure $d\mu$ by $d\mu(\theta) = \frac{1}{2\pi}P_r(\theta - t)\,d\theta$, then the above formula can be written as

$$f(re^{it}) = \int_0^{2\pi} f(e^{i\theta})\,d\mu(\theta).$$

Notice that $f(e^{i\theta}) \in \boldsymbol{L}^2(S^1, d\mu)$, since for fixed $r < 1$, $P_r(\theta - t)$ is bounded above. Applying the Cauchy–Schwarz inequality to the product of the functions $f(e^{i\theta}) \in \boldsymbol{L}^2(S^1, d\mu)$ and $1 \in \boldsymbol{L}^2(S^1, d\mu)$, we obtain

$$\begin{aligned}|f(re^{it})| &= \left|\int_0^{2\pi} f(e^{i\theta})\, d\mu(\theta)\right| \\ &\le \left(\int_0^{2\pi} |f(e^{i\theta})|^2\, d\mu(\theta)\right)^{1/2} \left(\int_0^{2\pi} |1|^2\, d\mu(\theta)\right)^{1/2} \\ &= \left(\int_0^{2\pi} |f(e^{i\theta})|^2\, d\mu(\theta)\right)^{1/2}\end{aligned}$$

since $\frac{1}{2\pi}\int_0^{2\pi} P_r(\theta - t)\, d\theta = 1$. Squaring both sides of the above inequality yields

$$|f(re^{it})|^2 \le \frac{1}{2\pi}\int_0^{2\pi} P_r(\theta - t)|f(e^{i\theta})|^2\, d\theta.$$

□

We can now prove that C_ϕ is a well-defined bounded operator on $\boldsymbol{H}^2$.

Theorem 5.1.5. *Let $\phi : \mathbb{D} \longrightarrow \mathbb{D}$ be analytic. Then the composition operator C_ϕ is well-defined and bounded on $\boldsymbol{H}^2$. Moreover,*

$$\|C_\phi\| \le \sqrt{\frac{1 + |\phi(0)|}{1 - |\phi(0)|}}.$$

Proof. Define the real-valued function u on $\mathbb{D}$ by

$$u(re^{it}) = \frac{1}{2\pi}\int_0^{2\pi} P_r(\theta - t)|f(e^{i\theta})|^2\, d\theta.$$

This is a harmonic function on $\mathbb{D}$, and, by the previous lemma,

$$|f(z)|^2 \le u(z),$$

for all $z \in \mathbb{D}$. Since the range of ϕ is in $\mathbb{D}$, by the inequality above,

$$|f(\phi(w))|^2 \le u(\phi(w))$$

for every $w \in \mathbb{D}$. Writing $w = re^{it}$ and integrating from 0 to 2π, we obtain

$$\frac{1}{2\pi}\int_0^{2\pi} |f(\phi(re^{it}))|^2\, dt \le \frac{1}{2\pi}\int_0^{2\pi} u(\phi(re^{it}))\, dt.$$

Since u is harmonic and ϕ is analytic, it follows that $u \circ \phi$ is also harmonic. Hence

$$u(\phi(0)) = \frac{1}{2\pi}\int_0^{2\pi} u(\phi(re^{it}))\, dt$$

by the mean value property of harmonic functions.

Therefore

$$\frac{1}{2\pi}\int_0^{2\pi} |f(\phi(re^{it}))|^2\, dt \le u(\phi(0)).$$

By Theorem 1.1.12, this implies that $f \circ \phi \in \boldsymbol{H}^2$; i.e., C_ϕ is well-defined.

To see that C_ϕ is bounded, observe that the last inequality implies (by Theorem 1.1.12) that

$$\|C_\phi f\|^2 \le u(\phi(0)).$$

Notice also that

$$P_r(\theta - t) = \frac{1-r^2}{1-2r\cos(\theta - t)+r^2} \le \frac{1-r^2}{(1-r)^2} = \frac{1+r}{1-r}.$$

Since

$$u(re^{it}) = \frac{1}{2\pi}\int_0^{2\pi} P_r(\theta - t)|f(e^{i\theta})|^2\, d\theta,$$

it follows that

$$u(re^{it}) \le \left(\frac{1+r}{1-r}\right)\frac{1}{2\pi}\int_0^{2\pi} |f(e^{i\theta})|^2\, d\theta = \left(\frac{1+r}{1-r}\right)\|f\|^2,$$

in other words,

$$u(z) \le \left(\frac{1+|z|}{1-|z|}\right)\|f\|^2$$

for every $z \in \mathbb{D}$.

In particular,

$$u(\phi(0)) \le \left(\frac{1+|\phi(0)|}{1-|\phi(0)|}\right)\|f\|^2.$$

Hence

$$\|C_\phi f\|^2 \le \left(\frac{1+|\phi(0)|}{1-|\phi(0)|}\right)\|f\|^2.$$

Therefore C_ϕ is bounded and

$$\|C_\phi\| \le \sqrt{\frac{1+|\phi(0)|}{1-|\phi(0)|}}\,.$$

□

The following observation will be useful. For any $\phi : \mathbb{D} \longrightarrow \mathbb{D}$ analytic, $C_\phi e_0 = e_0$. Thus $\|C_\phi\| \ge 1$ for every composition operator. This yields the following corollary.

Corollary 5.1.6. *If C_ϕ is a composition operator such that $\phi(0) = 0$ then $\|C_\phi\| = 1$.*

Proof. By the previous theorem and $\phi(0) = 0$, we have $\|C_\phi\| \leq 1$. By the above observation, $\|C_\phi\| \geq 1$. □

Composition operators are related to a classical concept in complex analysis called "subordination".

Definition 5.1.7. For functions f and g analytic on $\mathbb{D}$, f is *subordinate to* g if there exists an analytic function $\phi : \mathbb{D} \longrightarrow \mathbb{D}$ with $\phi(0) = 0$ such that $f = g \circ \phi$.

The previous corollary is a well-known classical theorem.

Corollary 5.1.8 (Littlewood's Subordination Theorem). *If f in $\boldsymbol{H}^2$ is subordinate to g in $\boldsymbol{H}^2$, then $\|f\| \leq \|g\|$.*

Proof. Apply the previous corollary to $f = C_\phi g$. □

Reproducing kernels give a lot of information about composition operators. Recall that (see Definition 1.1.7), for $\lambda \in \mathbb{D}$, the function k_λ defined by

$$k_\lambda(z) = \frac{1}{1 - \overline{\lambda} z}$$

has the property that $(f, k_\lambda) = f(\lambda)$ for every f in $\boldsymbol{H}^2$.

Lemma 5.1.9. *If C_ϕ is a composition operator and k_λ is a reproducing kernel function, then $C_\phi^* k_\lambda = k_{\phi(\lambda)}$.*

Proof. For each f in $\boldsymbol{H}^2$,

$$(f, C_\phi^* k_\lambda) = (C_\phi f, k_\lambda) = (f \circ \phi, k_\lambda) = f(\phi(\lambda)).$$

But also

$$(f, k_{\phi(\lambda)}) = f(\phi(\lambda)),$$

and therefore

$$(f, C^* k_\lambda) = (f, k_{\phi(\lambda)})$$

for all $f \in \boldsymbol{H}^2$. This implies that $C_\phi^* k_\lambda = k_{\phi(\lambda)}$. □

Theorem 5.1.10. *For every composition operator C_ϕ,*

$$\frac{1}{\sqrt{1 - |\phi(0)|^2}} \leq \|C_\phi\| \leq \frac{2}{\sqrt{1 - |\phi(0)|^2}}.$$

Proof. Using the previous lemma with $\lambda = 0$ yields

$$C_\phi^* k_0 = k_{\phi(0)}.$$

Recall (Theorem 1.1.8) that

$$\|k_\lambda\|^2 = \frac{1}{1 - |\lambda|^2},$$

and therefore $\|k_0\| = 1$ and $\|k_{\phi(0)}\| = \frac{1}{\sqrt{1-|\phi(0)|^2}}$. Since

$$\|k_{\phi(0)}\| = \|C_\phi^* k_0\| \leq \|C_\phi^*\| \, \|k_0\|,$$

it follows that

$$\frac{1}{\sqrt{1 - |\phi(0)|^2}} \leq \|C_\phi^*\| = \|C_\phi\|.$$

To prove the other inequality, we begin with the result from Theorem 5.1.5:

$$\|C_\phi\| \leq \sqrt{\frac{1 + |\phi(0)|}{1 - |\phi(0)|}}.$$

Observe that, for $0 \leq r < 1$, we have the inequality

$$\sqrt{\frac{1+r}{1-r}} = \sqrt{\frac{(1+r)^2}{1-r^2}} = \frac{1+r}{\sqrt{1-r^2}} \leq \frac{2}{\sqrt{1-r^2}}.$$

It follows that

$$\|C_\phi\| \leq \sqrt{\frac{1 + |\phi(0)|}{1 - |\phi(0)|}} \leq \frac{2}{\sqrt{1 - |\phi(0)|^2}}.$$

□

We showed (Corollary 5.1.6) that $\|C_\phi\| = 1$ if $\phi(0) = 0$. The converse is also true.

Corollary 5.1.11. *The norm of the composition operator C_ϕ is 1 if and only if $\phi(0) = 0$.*

Proof. As indicated, we have already established (Corollary 5.1.6) that $\|C_\phi\| = 1$ if $\phi(0) = 0$. Conversely, if $\|C_\phi\| = 1$, then the inequality

$$\frac{1}{\sqrt{1 - |\phi(0)|^2}} \leq \|C_\phi\|$$

implies

$$\frac{1}{\sqrt{1 - |\phi(0)|^2}} \leq 1,$$

so $\phi(0) = 0$. □

Composition operators are characterized as those operators whose adjoints map the set of reproducing kernels into itself.

Theorem 5.1.12. *An operator A on $\boldsymbol{H}^2$ is a composition operator if and only if A^* maps the set of reproducing kernels into itself.*

Proof. We showed above that $A^*k_\lambda = k_{\phi(\lambda)}$ when $A = C_\phi$. Conversely, suppose that for each $\lambda \in \mathbb{D}$, $A^*k_\lambda = k_{\lambda'}$ for some $\lambda' \in \mathbb{D}$. Define $\phi : \mathbb{D} \longrightarrow \mathbb{D}$ by $\phi(\lambda) = \lambda'$.

Notice that, for $f \in \boldsymbol{H}^2$,

$$(Af, k_\lambda) = (f, A^*k_\lambda) = (f, k_{\phi(\lambda)}) = f(\phi(\lambda)).$$

If we take $f(z) = z$, then $g = Af$ is in $\boldsymbol{H}^2$, and is thus analytic. But then, by the above equation, we have

$$g(\lambda) = (g, k_\lambda) = (Af, k_\lambda) = f(\phi(\lambda)) = \phi(\lambda).$$

Therefore $g = \phi$ and ϕ is analytic, so the composition operator C_ϕ is well-defined and bounded.

It follows that $A = C_\phi$, since

$$\begin{aligned}(Af)(\lambda) &= (Af, k_\lambda)\\ &= f(\phi(\lambda)) \qquad \text{(as shown above)}\\ &= (C_\phi f)(\lambda),\end{aligned}$$

for all f in $\boldsymbol{H}^2$. □

There are other interesting characterizations of composition operators. Recall that $e_n(z) = z^n$ for nonnegative integers n.

Theorem 5.1.13. *An operator A in $\boldsymbol{H}^2$ is a composition operator if and only if $Ae_n = (Ae_1)^n$ for $n = 0, 1, 2, \ldots$.*

Proof. If $A = C_\phi$, then $Ae_1 = C_\phi e_1 = C_\phi z = \phi$ and $Ae_n = C_\phi e_n = C_\phi z^n = \phi^n$, and therefore $(Ae_1)^n = Ae_n$.

Conversely, suppose $Ae_n = (Ae_1)^n$ for all nonnegative integers n. Define ϕ by $\phi = Ae_1$. Since Ae_1 is in $\boldsymbol{H}^2$, ϕ is analytic on $\mathbb{D}$.

To show that $A = C_\phi$, it suffices to prove that $|\phi(z)| < 1$ for all $z \in \mathbb{D}$, since then it would follow that the composition operator C_ϕ is well-defined and bounded. Then

$$Ae_n = (Ae_1)^n = \phi^n = C_\phi z^n = C_\phi e_n;$$

thus by linearity and continuity, it would follow that $A = C_\phi$.

To show that $|\phi(z)| < 1$, note that $\phi^n = Ae_n$ implies that $\|\phi^n\| \leq \|A\|$ for all nonnegative n. We claim that $|\widetilde{\phi}(e^{i\theta})| \leq 1$ for almost all θ. Consider any $\delta > 0$ and define the set E by $E = \{e^{i\theta} : |\widetilde{\phi}(e^{i\theta})| \geq 1 + \delta\}$. Then

$$\begin{aligned}\|\phi^n\|^2 &= \frac{1}{2\pi}\int_0^{2\pi} |\widetilde{\phi}(e^{i\theta})|^{2n}\, d\theta \\ &\geq \frac{1}{2\pi}\int_E |\widetilde{\phi}(e^{i\theta})|^{2n}\, d\theta \\ &\geq \int_E (1+\delta)^{2n}\ dm \\ &= m(E)(1+\delta)^{2n}\end{aligned}$$

where $m(E)$ is the measure of E. If $m(E) > 0$, this would imply that $\{\|\phi^n\|\} \to \infty$ as $n \to \infty$ which contradicts the fact that $\|\phi^n\| \leq \|A\|$ for all n. Hence $m(E) = 0$ and therefore $|\widetilde{\phi}(e^{i\theta})| \leq 1$. It follows that $|\phi(z)| \leq 1$ for all z in $\mathbb{D}$ (Corollary 1.1.24).

We claim that $|\phi(z)| < 1$ for all z in $\mathbb{D}$. If not, then there exists $z_0 \in \mathbb{D}$ such that $|\phi(z_0)| = 1$. By the maximum modulus principle ([9, pp. 79, 128], [47, p. 212]) this implies that ϕ is a constant function; say $\phi(z) = \lambda$, with λ of modulus 1. Since $Ae_n = (Ae_1)^n$, it follows that $Ae_n = \lambda^n$. But then $(A^*e_0, e_n) = (e_0, Ae_n) = (e_0, \lambda^n) = \overline{\lambda}^n$, so

$$\|A^*e_0\|^2 = \sum_{k=0}^{\infty} |(A^*e_0, e_n)|^2 = \sum_{k=0}^{\infty} |\lambda|^{2n} = \infty,$$

since $|\lambda| = 1$. This is a contradiction. □

Corollary 5.1.14. *The operator A on $\boldsymbol{H}^2$ is a composition operator if and only if it is multiplicative in the sense that $(Af)(Ag) = A(fg)$ whenever f, g, and fg are all in $\boldsymbol{H}^2$.*

Proof. It is clear that composition operators have the stated multiplicative property.

Conversely, if A has the multiplicative property then, in particular, $Ae_n = (Ae_1)^n$ for all n, so the fact that A is a composition operator follows from Theorem 5.1.13. □

There are very few normal composition operators.

Theorem 5.1.15. *The composition operator C_ϕ is normal if and only if there exists $\lambda \in \mathbb{C}$ such that $\phi(z) = \lambda z$ and $|\lambda| \leq 1$.*

Proof. First note that $\phi(z) = \lambda z$ implies that $C_\phi e_n = \lambda^n e_n$ for all positive integers n. Hence C_ϕ is a diagonal operator with respect to the canonical basis of $\boldsymbol{H}^2$, and it is obvious that every diagonal operator is normal.

To establish the converse, suppose C_ϕ is normal. Note that, for every complex number α, the operator $C_\phi - \alpha$ is also normal, so

$$\|(C_\phi - \alpha)f\| = \|(C_\phi - \alpha)^* f\|$$

holds for every $f \in \boldsymbol{H}^2$. In particular, $C_\phi^* f = \overline{\alpha} f$ if and only if $C_\phi f = \alpha f$.

Since $C_\phi e_0 = e_0$ for every C_ϕ, the above implies that $C_\phi^* e_0 = e_0$ when C_ϕ is normal. But e_0 is the kernel function k_0, and, for every composition operator, $C_\phi^* k_\lambda = k_{\phi(\lambda)}$ (by Lemma 5.1.9). It follows that $k_{\phi(0)} = k_0$, so $\phi(0) = 0$.

Since $\phi(0) = 0$, it follows that the subspace $z^2\boldsymbol{H}^2$ is invariant under C_ϕ. (Clearly, $C_\phi(z^2 g(z))$ has a zero of multiplicity at least 2 at 0 whenever $\phi(0) = 0$ and g is in $\boldsymbol{H}^2$.) Thus C_ϕ^* leaves the subspace $\left(z^2\boldsymbol{H}^2\right)^\perp = \bigvee\{e_0, e_1\}$ invariant (Theorem 1.2.20). Since $\bigvee\{e_0\}$ is an invariant subspace of C_ϕ, it also follows that $z\boldsymbol{H}^2 = \left(\bigvee\{e_0\}\right)^\perp$ is an invariant subspace of C_ϕ^*.

Thus C_ϕ^* leaves invariant the subspaces $z\boldsymbol{H}^2$ and $\bigvee\{e_0, e_1\}$. Therefore $z\boldsymbol{H}^2 \cap \bigvee\{e_0, e_1\} = \bigvee\{e_1\}$ is also an invariant subspace for C_ϕ^*. This yields $C_\phi^* e_1 = \overline{\lambda} e_1$ for some $\lambda \in \mathbb{C}$. Since C_ϕ is normal, it follows that $C_\phi e_1 = \lambda e_1$. Obviously, $C_\phi e_1 = \phi$, so $\phi(z) = \lambda z$.

That $|\lambda| \leq 1$ is obvious since $\phi(z) \in \mathbb{D}$ for all $z \in \mathbb{D}$. □

There are a fair number of compact composition operators.

Theorem 5.1.16. *If there exists a positive number $s < 1$ so that $|\phi(z)| < s$ for every $z \in \mathbb{D}$, then C_ϕ is compact.*

Proof. We show that C_ϕ is compact by exhibiting a sequence of operators of finite rank that converge in norm to C_ϕ. Observe that, since $|\phi(z)| < s$ for all $z \in \mathbb{D}$, we have, for each natural number k,

$$\|\phi^k\| \leq \|\phi^k\|_\infty \leq \|\phi\|_\infty^k \leq s^k.$$

For every natural number n define the finite-rank operator F_n by

$$F_n = \sum_{k=0}^{n} \phi^k \otimes e_k.$$

If g is in $\boldsymbol{H}^2$, then, since $s < 1$,

$$\begin{aligned}\|(F_n - C_\phi)g\| &= \left\| \sum_{k=0}^{n} (g, e_k)\,\phi^k - \sum_{k=0}^{\infty} (g, e_k)\,\phi^k \right\| \\ &= \left\| \sum_{k=n+1}^{\infty} (g, e_k)\,\phi^k \right\| \\ &\le \sum_{k=n+1}^{\infty} |(g, e_k)|\, \|\phi^k\| \\ &\le \sum_{k=n+1}^{\infty} |(g, e_k)|\, s^k \\ &\le \left(\sum_{k=n+1}^{\infty} |(g, e_k)|^2 \right)^{1/2} \left(\sum_{k=n+1}^{\infty} s^{2k} \right)^{1/2} \\ &\le \|g\| \frac{s^{n+1}}{\sqrt{1 - s^2}},\end{aligned}$$

where we have used the Cauchy–Schwarz inequality. This implies that

$$\|F_n - C_\phi\| \le \frac{s^{n+1}}{\sqrt{1 - s^2}}$$

and thus that $\{F_n\} \to C_\phi$ uniformly. □

For every analytic ϕ taking $\mathbb{D}$ into $\mathbb{D}$ that is not constant, $|\phi(z)| < 1$ for all $z \in \mathbb{D}$ by the maximum modulus theorem ([9, pp. 79, 128], [47, p. 212]). If C_ϕ is compact, a stronger condition holds.

Theorem 5.1.17. *If C_ϕ is compact, then $|\widetilde{\phi}(e^{i\theta})| < 1$ a.e.*

Proof. The sequence $\{e_n\}$ converges weakly to 0. So, $\{C_\phi e_n\}$ converges to 0 in norm if C_ϕ is compact (e.g., [27, p. 95] or [12, p. 173]).

If $|\widetilde{\phi}(e^{i\theta})|$ was not less than 1 a.e., there would exist a subset E of the unit circle, with positive measure, such that $|\widetilde{\phi}(e^{i\theta})| = 1$ for $e^{i\theta} \in E$. Then, for each positive integer n,

$$\|\widetilde{\phi}^n\|^2 = \frac{1}{2\pi} \int_0^{2\pi} |\widetilde{\phi}(e^{i\theta})|^{2n}\, d\theta$$

is greater than or equal to

$$\frac{1}{2\pi} \int_E |\widetilde{\phi}(e^{i\theta})|^{2n}\, d\theta = m(E).$$

This contradicts the fact that $\{C_\phi e_n\} = \{\phi^n\}$ converges to 0 in norm. □

It is clear that a necessary and sufficient condition that C_ϕ be compact should involve a condition on ϕ in between those of the previous two theorems. Such a condition is not at all easy to find. However, Shapiro ([53, Chapter 10] and [157]) did obtain a beautiful characterization of compact composition operators in terms of the Nevanlinna counting function of ϕ.

5.2 Invertibility of Composition Operators

Recall that the conformal mappings of $\mathbb{D}$ onto itself are the functions of the form

$$\phi(z) = e^{i\theta}\frac{\lambda - z}{1 - \overline{\lambda}z},$$

for λ a fixed element of $\mathbb{D}$ and θ a fixed real number ([9, p. 132], [47, p. 255]).

Theorem 5.2.1. *The composition operator C_ϕ is invertible if and only if ϕ is a conformal mapping of $\mathbb{D}$ onto itself. In this case, $C_\phi^{-1} = C_{\phi^{-1}}$.*

Proof. If ϕ is a conformal map, let ϕ^{-1} be the inverse conformal map. Then $C_{\phi^{-1}}C_\phi = C_\phi C_{\phi^{-1}} = I$ by Theorem 5.1.2. Hence $C_\phi^{-1} = C_{\phi^{-1}}$.

To establish the converse, suppose that C_ϕ is an invertible composition operator; let A denote its inverse.

Define ψ to be Ae_1. Our goal is to prove that $A = C_{\phi^{-1}}$; we begin by proving that A is a composition operator. By Theorem 5.1.13, this will follow if we establish that Ae_n is $(Ae_1)^n = \psi^n$. For any fixed n, define $g = Ae_n$. Since A is the inverse of C_ϕ, it follows that

$$e_n = C_\phi A e_n = C_\phi g = g \circ \phi.$$

On the other hand,

$$e_1 = C_\phi A e_1 = C_\phi \psi = \psi \circ \phi.$$

Since $e_1^n = e_n$, we have

$$(\psi \circ \phi)^n = g \circ \phi.$$

Note that ϕ cannot be a constant function (since the range of C_ϕ would then consist of constant functions, so C_ϕ would not be onto). By the open mapping theorem of complex analysis, $\phi(\mathbb{D})$ is an open subset of $\mathbb{D}$. This means that the equation

$$(\psi(w))^n = g(w)$$

holds for all w in the open set $\phi(\mathbb{D})$, and therefore, by analyticity, it holds for all $w \in \mathbb{D}$. Thus $g = \psi^n$; i.e., $Ae_n = (Ae_1)^n$, as desired.

Since $A = C_\psi$ is a composition operator and $C_\psi C_\phi = C_\phi C_\psi = I$, we then have

$$(\phi \circ \psi)(z) = z \quad \text{and} \quad (\psi \circ \phi)(z) = z,$$

which implies that ϕ is a conformal map of $\mathbb{D}$ into itself. □

Example 5.2.2. *For a fixed $\lambda \in \mathbb{D}$, define the function ϕ_λ by*

$$\phi_\lambda(z) = \frac{\lambda - z}{1 - \overline{\lambda} z}.$$

Then $(C_{\phi_\lambda})^2 = I$ (i.e., $C_{\phi_\lambda}^{-1} = C_{\phi_\lambda}$).

Proof. As indicated above, each such ϕ_λ is a conformal mapping of $\mathbb{D}$ into itself. An easy computation shows that $\phi_\lambda(\phi_\lambda(z)) = z$ for all $z \in \mathbb{D}$, from which the result follows. □

Theorem 5.2.3. *If the function ϕ has a fixed point in $\mathbb{D}$, then the operator C_ϕ is similar to a composition operator C_ψ with the property that $\psi(0) = 0$.*

Proof. Let $\phi(\lambda) = \lambda$ for some $\lambda \in \mathbb{D}$. Let

$$\phi_\lambda(z) = \frac{\lambda - z}{1 - \overline{\lambda} z}.$$

Then $C_{\phi_\lambda}^{-1} = C_{\phi_\lambda}$ by Example 5.2.2, so

$$C_{\phi_\lambda}^{-1} C_\phi C_{\phi_\lambda} = C_{\phi_\lambda \circ \phi \circ \phi_\lambda} \qquad \text{(by Theorem 5.1.2)}$$

and $(\phi_\lambda \circ \phi \circ \phi_\lambda)(0) = 0$. □

A special case of $C_\phi C_\psi = C_{\psi \circ \phi}$ is the assertion that, for each natural number n, C_ϕ^n is the composition operator induced by the function obtained by composing ϕ with itself n times. It will be useful to have notation for such functions.

Notation 5.2.4. For any function $\phi : \mathbb{D} \longrightarrow \mathbb{D}$ and positive integer n, we define the function $\phi^{[n]} : \mathbb{D} \longrightarrow \mathbb{D}$ recursively by $\phi^{[1]}(z) = \phi(z)$ and $\phi^{[n]}(z) = (\phi \circ \phi^{[n-1]})(z)$.

Some information about spectra of composition operators is easily obtainable in the cases in which the inducing function has a fixed point.

Theorem 5.2.5. *If ϕ has a fixed point in $\mathbb{D}$, then the spectral radius of C_ϕ is* 1.

Proof. By Theorem 5.2.3, C_ϕ is similar to a composition operator whose defining function fixes the point 0. Since similar operators have the same spectra, and therefore equal spectral radii, we may and do assume that $\phi(0) = 0$.

By the spectral radius formula (Theorem 1.2.4),

$$r(C_\phi) = \lim_{n\to\infty} \|C_\phi^n\|^{1/n}.$$

For every n, $C_\phi^n = C_{\phi^{[n]}}$ (cf. Notation 5.2.4). Since $\phi^{[n]}(0) = 0$, the norm of $C_{\phi^{[n]}}$ equals 1 (by Corollary 5.1.11). Hence $\|C_\phi^n\| = 1$. Then

$$r(C_\phi) = \lim_{n\to\infty} \|C_\phi^n\|^{1/n} = 1.$$

□

5.3 Eigenvalues and Eigenvectors

We begin with a result about point spectra of the adjoints of those composition operators whose inducing functions have fixed points on the disk.

Theorem 5.3.1. *If C_ϕ is a composition operator such that $\phi(a) = a$ for some $a \in \mathbb{D}$, then*

$$\Pi_0(C_\phi^*) \supset \{1\} \cup \bigcup_{k=1}^{\infty} \left\{ \left(\overline{\phi'(a)}\right)^k \right\}.$$

Proof. If $\phi_a(z) = \frac{a-z}{1-\overline{a}z}$ then $C_{\phi_a} C_\phi C_{\phi_a} = C_\psi$, where $\psi = \phi_a \circ \phi \circ \phi_a$, by Theorem 5.1.2. Since similarity preserves point spectra, C_ψ has the same point spectrum as C_ϕ. Applying the chain rule to $\psi = \phi_a \circ \phi \circ \phi_a$ yields

$$\psi'(z) = \phi_a'(\phi(\phi_a(z)))\ \phi'(\phi_a(z))\ \phi_a'(z).$$

In particular,

$$\psi'(0) = \phi_a'(a)\phi'(a)\phi_a'(0),$$

which an easy computation shows is equal to $\phi'(a)$. Thus it suffices to prove the theorem in the case $a = 0$.

Suppose, then, that $\phi(0) = 0$.

Fix a natural number n. For every $g \in \boldsymbol{H}^2$, $C_\phi(z^n g(z)) = (\phi(z))^n g(\phi(z))$, so $\phi(0) = 0$ implies that $C_\phi(z^n g(z))$ has a zero of multiplicity at least n at 0. Thus, $\phi(0) = 0$ implies that the subspace $z^n \boldsymbol{H}^2$ is invariant under C_ϕ.

It follows (Theorem 1.2.20) that its orthocomplement, $\bigvee\{e_0, e_1, \ldots, e_{n-1}\}$, is invariant under C_ϕ^*.

Therefore, for each natural number n, the restriction of C_ϕ^* to $\bigvee\{e_0, e_1, \ldots, e_{n-1}\}$ has a matrix with respect to the basis $\{e_0, e_1, \ldots, e_{n-1}\}$ that is upper triangular. As is very well known and easy to verify, the diagonal entries of an upper triangular matrix are eigenvalues of the matrix. The diagonal entries in the present case are $\{(C_\phi^* e_k, e_k) : k = 0, 1, \ldots, n-1\}$.

The case $k = 0$ gives $(C_\phi^* e_0, e_0) = (e_0, e_0) = 1$. For $k \geq 1$,

$$(C_\phi^* e_k, e_k) = (e_k, C_\phi e_k) = (e_k, \phi^k).$$

Since $\phi(0) = 0$, the power series representation of ϕ has the form

$$\phi(z) = \phi'(0)z + \frac{\phi''(0)}{2!}z^2 + \frac{\phi'''(0)}{3!}z^3 + \cdots.$$

Hence the coefficient of the term of degree k of the power series of ϕ^k is $(\phi'(0))^k$. Therefore

$$(C_\phi^* e_k, e_k) = (e_k, \phi^k)$$

gives

$$(C_\phi^* e_k, e_k) = \overline{\phi'(0)}^k.$$

Thus the restriction of C_ϕ^* to $\bigvee\{e_0, e_1, \ldots, e_{n-1}\}$ has eigenvalues

$$\{1\} \bigcup \left\{ \overline{\phi'(0)}^k : k = 1, 2, \ldots, n-1 \right\}.$$

An eigenvalue of a restriction of an operator to an invariant subspace is obviously an eigenvalue of the operator itself (corresponding to the same eigenvector). Hence the above set is contained in $\Pi_0(C_\phi^*)$. Since this is true for every n, the theorem is established. □

As the unilateral shift and many other examples indicate, the adjoint of an operator can have eigenvalues even when the operator itself does not have any. Thus the above result does not yield any information about any eigenvalues C_ϕ may have. (A result limiting the eigenvalues for C_ϕ is given in Corollary 5.3.4 for those ϕ that have a fixed point in $\mathbb{D}$.) It does, however, give some information about $\sigma(C_\phi)$.

Corollary 5.3.2. *If C_ϕ is a composition operator and $\phi(a) = a$ for some $a \in \mathbb{D}$, then*

$$\sigma(C_\phi) \supset \{1\} \cup \bigcup_{k=1}^{\infty} \left\{ (\phi'(a))^k \right\}.$$

Proof. This follows immediately from the previous theorem and Theorem 1.2.4. □

Eigenfunctions of the composition operator C_ϕ are solutions of an equation of the form

$$f(\phi(z)) = \lambda f(z);$$

such equations are known as *Schröder* equations, and were studied for general analytic functions for many years before the Hardy–Hilbert space was defined.

In particular, in 1884, Königs [108] obtained a great deal of information about solutions of the Schröder equation. The following very simple result suffices to determine the point spectrum of many composition operators.

Theorem 5.3.3. *If ϕ is a nonconstant analytic function mapping the disk into itself and satisfying $\phi(a) = a$ for some $a \in \mathbb{D}$, and if there exists a function f analytic on $\mathbb{D}$ that is not identically zero and satisfies the Schröder equation*

$$f(\phi(z)) = \lambda f(z)$$

for some λ, then either $\lambda = 1$ or there is a natural number k such that $\lambda = (\phi'(a))^k$.

Proof. The equations $\phi(a) = a$ and $f(\phi(z)) = \lambda f(z)$ yield $f(a) = \lambda f(a)$. If $f(a) \neq 0$, then clearly $\lambda = 1$ and the theorem is established in that case.

Suppose $f(a) = 0$. Since f is not identically zero, $f(z)$ has a power series expansion

$$f(z) = b_k(z-a)^k + b_{k+1}(z-a)^{k+1} + \cdots,$$

with $b_k \neq 0$ for some $k \geq 1$.

It follows that

$$\frac{f(\phi(z))}{f(z)} = \left(\frac{\phi(z)-a}{z-a}\right)^k \left(\frac{b_k + b_{k+1}(\phi(z)-a) + b_{k+2}(\phi(z)-a)^2 + \cdots}{b_k + b_{k+1}(z-a) + +b_{k+2}(z-a)^2 + \cdots}\right).$$

Since $\phi(a) = a$,

$$\lim_{z \to a} \frac{\phi(z)-a}{z-a} = \phi'(a).$$

Also,

$$\lim_{z \to a} \left(b_k + b_{k+1}(\phi(z)-a) + b_{k+2}(\phi(z)-a)^2 + \cdots\right) = b_k$$

and

$$\lim_{z\to a}\left(b_k + b_{k+1}(z-a) + +b_{k+2}(z-a)^2 + \cdots\right) = b_k.$$

Therefore

$$\begin{aligned}\lim_{z\to a}\frac{f(\phi(z))}{f(z)} &= \lim_{z\to a}\left(\frac{\phi(z)-a}{z-a}\right)^k\left(\frac{b_k + b_{k+1}(\phi(z)-a) + b_{k+2}(\phi(z)-a)^2 + \cdots}{b_k + b_{k+1}(z-a) + +b_{k+2}(z-a)^2 + \cdots}\right)\\ &= (\phi'(a))^k \cdot 1\\ &= (\phi'(a))^k.\end{aligned}$$

However, $\frac{f(\phi(z))}{f(z)}$ is identically λ, so $\lambda = (\phi'(a))^k$. □

Corollary 5.3.4. *If C_ϕ is a composition operator and $\phi(a) = a$ for some $a \in \mathbb{D}$, then*

$$\Pi_0(C_\phi) \subset \{1\} \cup \bigcup_{k=1}^{\infty}\{(\phi'(a))^k\} \subset \sigma(C_\phi).$$

Proof. This corollary is an immediate consequence of Corollary 5.3.2 and Theorem 5.3.3. □

The well-known Fredholm alternative ([12, p. 217], [27, p. 96]) is equivalent to the statement that every point other than 0 in the spectrum of a compact operator is an eigenvalue.

Corollary 5.3.5. *If C_ϕ is a compact composition operator satisfying $\phi(a) = a$ for some $a \in \mathbb{D}$, then*

$$\sigma(C_\phi) = \{0\} \cup \{1\} \cup \bigcup_{k=1}^{\infty}\{(\phi'(a))^k\}.$$

Proof. This follows immediately from the previous corollary and the Fredholm alternative. □

5.4 Composition Operators Induced by Disk Automorphisms

We have seen that the composition operator C_ϕ is invertible if and only if ϕ is a conformal mapping of $\mathbb{D}$ onto itself (Theorem 5.2.1).

Recall that every conformal mapping of $\mathbb{D}$ onto itself has the form

$$\phi(z) = \lambda\frac{a-z}{1-\overline{a}z},$$

where a is in $\mathbb{D}$ and λ has modulus 1. Such functions are generally called "disk automorphisms". In particular, each disk automorphism is a linear fractional transformation (or Möbius transformation); i.e., a function of the form

$$\phi(z) = \frac{az+b}{cz+d}$$

for complex numbers a, b, c, and d.

There are a number of interesting composition operators induced by disk automorphisms. It is customary to classify disk automorphisms by the nature of their fixed points, as in Definition 5.4.2 below.

It is clear that each disk automorphism is a continuous mapping of the closure of $\mathbb{D}$ into itself. Thus Brouwer's fixed-point theorem [12, p. 149] implies that each disk automorphism fixes at least one point in the closure of $\mathbb{D}$; this will be shown directly in the proof of the next theorem.

Theorem 5.4.1. *Let ϕ be an automorphism of the unit disk other than the identity automorphism. Then ϕ has at most two fixed points in the complex plane. Either ϕ has one fixed point inside $\mathbb{D}$ and does not fix any point in S^1 or ϕ has all of its fixed points in S^1.*

Proof. Note that the equation $\phi(z) = z$ is either a linear or a quadratic equation and thus has at most two roots. Hence ϕ has at most two fixed points in the complex plane.

Let $\phi(z) = \lambda \frac{a-z}{1-\overline{a}z}$. If $\phi(0) = 0$, then $a = 0$ and the only fixed point of ϕ is 0. If $\phi(z_0) = z_0$ and $z_0 \neq 0$, then

$$\phi\left(\frac{1}{\overline{z_0}}\right) = \lambda\left(\frac{a - \frac{1}{\overline{z_0}}}{1 - \overline{a}\frac{1}{\overline{z_0}}}\right) = \lambda\left(\frac{1 - a\overline{z_0}}{\overline{a} - \overline{z_0}}\right) = \overline{\left(\frac{1}{\lambda\left(\frac{a-z_0}{1-\overline{a}z_0}\right)}\right)} = \overline{\left(\frac{1}{z_0}\right)} = \frac{1}{\overline{z_0}}.$$

Therefore $\frac{1}{\overline{z_0}}$ is also a fixed point in this case. Thus if ϕ has a fixed point other than 0 inside $\mathbb{D}$, its other fixed point is outside the closure of $\mathbb{D}$. The theorem follows. □

Definition 5.4.2. Let ϕ be an automorphism of $\mathbb{D}$ other than the identity transformation. If ϕ has a fixed point in $\mathbb{D}$, then ϕ is said to be *elliptic.* If ϕ does not have a fixed point in $\mathbb{D}$ and has only one fixed point in S^1, then ϕ is said to be *parabolic.* If ϕ has two fixed points in S^1, then ϕ is said to be *hyperbolic.*

It is very easy to find the spectrum of C_ϕ in the case that ϕ is an elliptic disk automorphism.

Theorem 5.4.3. *If ϕ is a disk automorphism satisfying $\phi(a) = a$ for some $a \in \mathbb{D}$, then the spectrum of C_ϕ is the closure of*

$$\{(\phi'(a))^n \ : \ n = 0, 1, 2, \ldots\}.$$

Proof. Recall that if $\psi = \phi_a \circ \phi \circ \phi_a$, then $\psi(0) = 0$ (see the proof of Theorem 5.2.3). Now, ψ is a disk automorphism so it has the form $\psi(z) = \mu \frac{b-z}{1-\bar{b}z}$ for some b in $\mathbb{D}$ and μ of modulus 1. But $\psi(0) = 0$, so $b = 0$; thus $\psi(z) = -\mu z$. Letting $\lambda = -\mu$ we have $\psi(z) = \lambda z$ and $|\lambda| = 1$.

As we have already observed (see the proof of Theorem 5.1.15), the operator C_ψ has a diagonal matrix with respect to the standard basis of $\boldsymbol{H}^2$; the diagonal entries of that matrix are $\{1, \lambda, \lambda^2, \lambda^3, \ldots\}$. As is well known and easy to prove, the spectrum of a diagonal matrix is the closure of the set of diagonal entries. Thus the spectrum of C_ψ is the closure of $\{1, \lambda, \lambda^2, \lambda^3, \ldots\}$. Note that $\lambda = \psi'(0)$.

Recall that $C_\phi = C_{\phi_a} C_\psi C_{\phi_a}$. Since similar operators have the same spectra, the spectrum of C_ϕ is the closure of $\{1, \lambda, \lambda^2, \lambda^3, \ldots\}$. Since $\lambda = \psi'(0) = \phi'(a)$, the theorem follows. □

The case of parabolic disk automorphisms is more interesting and more difficult than the elliptic case.

We can determine the spectral radius of C_ϕ by the spectral radius formula (Theorem 1.2.4) if we can compute the norm of C_ϕ^n. It turns out to be much easier to compute $\phi^{[n]}$ if we consider the iterates of a corresponding automorphism of the upper half-plane. It suffices to do so in the special case that the fixed point of ϕ is 1.

Lemma 5.4.4. *Let ϕ be a parabolic disk automorphism satisfying $\phi(1) = 1$. If $\Gamma(z) = i\frac{1+z}{1-z}$ and the function F is defined by $F = \Gamma \circ \phi \circ \Gamma^{-1}$, then there is a real number β such that $F(z) = z + \beta$ for all z in the open upper half-plane.*

Proof. Note that the function Γ is a conformal mapping of the open unit disk onto the open upper half-plane. Moreover, Γ maps S^1 bijectively onto $\mathbb{R} \cup \{\infty\}$. Notice that $\Gamma(1) = \infty$. Clearly, F is an automorphism of the upper half-plane and $F(\infty) = \infty$. Since the composition of linear fractional transformations is also a linear fractional transformation, F is linear a fractional transformation; thus F must be of the form

$$F(z) = \frac{az + b}{cz + d}$$

for some complex numbers a, b, c, and d. Since $F(\infty) = \infty$, it follows that $c = 0$. Thus we may write F in the form

$$F(z) = \alpha z + \beta$$

for suitable complex numbers α and β. Since ϕ is an automorphism of $\mathbb{D}$, it must map S^1 bijectively to S^1, and therefore F maps the extended real numbers onto the extended real numbers. Since $F(0) = \beta$, it follows that β must be a real number.

We claim that α is real. For suppose $\operatorname{Im} \alpha < 0$. Then $\overline{\alpha}$ is in the upper half-plane and $F(\overline{\alpha}) = |\alpha|^2 + \beta$ is real, contradicting the fact that F maps the open upper half-plane onto itself. Similarly, if $\operatorname{Im} \alpha > 0$ then $F(1) = \alpha + \beta$, which is in the open upper half-plane. This would contradict the fact that $F(1)$ is real. Therefore α is real.

Now, ∞ is the only fixed point that F has in the extended closed upper half-plane since, if F had another fixed point, it would follow that ϕ has at least two fixed points in the closure of $\mathbb{D}$. This implies that $\alpha = 1$, for if α was different from 1 then $\beta/(1-\alpha)$ would be another fixed point of F in the closed upper half-plane. Therefore F has the required form. □

Theorem 5.4.5. *If ϕ is a parabolic disk automorphism, then $\sigma(C_\phi) \subset S^1$; in particular, $r(C_\phi) = 1$.*

Proof. We will first determine the spectral radius by the spectral radius formula (Theorem 1.2.4). For this it is necessary to at least estimate the norm of C_ϕ^n. To use the preceding lemma, we must reduce the proof to the case that the fixed point is 1.

Assume that $\phi(e^{i\theta_0}) = e^{i\theta_0}$ for some real number θ_0. If we define T by $T(z) = e^{i\theta_0} z$, and ψ by $\psi = T^{-1} \circ \phi \circ T$, then ψ is also a parabolic disk automorphism, and $\psi(1) = 1$. As in the elliptic case, C_ψ is similar to C_ϕ and therefore they have the same spectra.

Thus we may and do assume from now on that $\phi(1) = 1$.

By the above lemma (Lemma 5.4.4), the function $F = \Gamma \circ \phi \circ \Gamma^{-1}$ has the form $F(z) = z + \beta$ for a real number β. Hence

$$F^{[n]}(z) = z + n\beta.$$

Recall that

$$\|C_\phi^n\| \leq \sqrt{\frac{1 + |\phi^{[n]}(0)|}{1 - |\phi^{[n]}(0)|}}.$$

Since clearly $\phi^{[n]} = \Gamma^{-1} \circ F^{[n]} \circ \Gamma$, it follows that

$$\phi^{[n]}(0) = \left(\Gamma^{-1} \circ F^{[n]} \circ \Gamma\right)(0) = \Gamma^{-1}(F^{[n]}(i)) = \Gamma^{-1}(i + n\beta).$$

It is easily seen that

$$\Gamma^{-1}(w) = \frac{w-i}{w+i}$$

for all w in the upper half-plane. Hence

$$\phi^{[n]}(0) = \frac{i + n\beta - i}{i + n\beta + i} = \frac{n\beta}{2i + n\beta}.$$

Therefore

$$\|C_\phi^n\| \leq \sqrt{\frac{|2i + n\beta| + |n\beta|}{|2i + n\beta| - |n\beta|}}.$$

But

$$\sqrt{\frac{|2i + n\beta| + |n\beta|}{|2i + n\beta| - |n\beta|}} = \sqrt{\frac{(|2i + n\beta| + |n\beta|)^2}{|2i + n\beta|^2 - |n\beta|^2}} = \frac{|2i + n\beta| + |n\beta|}{2}.$$

Applying the triangle inequality yields

$$\frac{|2i + n\beta| + |n\beta|}{2} \leq \frac{2 + n|\beta| + n|\beta|}{2} = 1 + n|\beta|.$$

Therefore

$$\|C_\phi^n\| \leq 1 + n|\beta|$$

for every positive integer n. It follows that

$$\lim_{n\to\infty} \|C_\phi^n\|^{1/n} \leq \lim_{n\to\infty} (1 + n|\beta|)^{1/n}.$$

But this last limit is easily seen to be 1, so the spectral radius formula (Theorem 1.2.4) implies that $r(C_\phi) \leq 1$.

We can establish that $r(C_\phi) = 1$ by applying the above to ϕ^{-1}; that is, since the inverse of a disk automorphism is a disk automorphism, ϕ^{-1} is a disk automorphism. Moreover, ϕ and ϕ^{-1} have the same fixed points, so ϕ^{-1} is parabolic. Thus, by the above, $r(C_\phi^{-1}) \leq 1$.

It follows easily that $\sigma(C_\phi)$ is contained in S^1. For suppose that $\lambda \in \sigma(C_\phi)$ and $|\lambda| < 1$. Then $1/\lambda$ is in $\sigma(C_\phi^{-1})$ by Theorem 1.2.4. Since $|1/\lambda| > 1$ this would contradict $r(C_\phi^{-1}) \leq 1$. Hence $\sigma(C_\phi) \subset S^1$ and $r(C_\phi) = 1$. □

The above theorem can be considerably sharpened.

Theorem 5.4.6. *If ϕ is a parabolic disk automorphism, then*

$$\sigma(C_\phi) = \Pi_0(C_\phi) = S^1.$$

Proof. By the previous theorem, it suffices to show that $S^1 \subset \Pi_0(C_\phi)$. As in the proof of the previous theorem, we can and do assume that $\phi(1) = 1$ and define Γ and F as in Lemma 5.4.4. In particular, $F(z) = z + \beta$ for some real β.

Fix λ in S^1. We can represent λ in the form $e^{i\theta_0}$, where θ_0 is a real number with the same sign as β. Define f_λ by

$$f_\lambda(z) = \exp\left(i\frac{\theta_0}{\beta}\Gamma(z)\right).$$

We will show that $C_\phi f_\lambda = \lambda f_\lambda$. We must first show that f_λ is in $\boldsymbol{H}^2$. Notice that

$$\begin{aligned} f_\lambda(z) &= \exp\left(i\frac{\theta_0}{\beta}\Gamma(z)\right) \\ &= \exp\left(i\frac{\theta_0}{\beta}\left(\operatorname{Re}\Gamma(z) + i\operatorname{Im}\Gamma(z)\right)\right) \\ &= \exp\left(-\frac{\theta_0}{\beta}\operatorname{Im}\Gamma(z)\right)\exp\left(i\left(\frac{\theta_0}{\beta}\operatorname{Re}\Gamma(z)\right)\right), \end{aligned}$$

and thus

$$|f_\lambda(z)| = \exp\left(-\frac{\theta_0}{\beta}\operatorname{Im}\Gamma(z)\right).$$

Recall that Γ maps $\mathbb{D}$ onto the upper half-plane, and thus, since β and θ_0 have the same sign, it follows that

$$-\frac{\theta_0}{\beta}\operatorname{Im}\Gamma(z) < 0.$$

Therefore $|f_\lambda(z)| < 1$ for all $z \in \mathbb{D}$. Thus each such f_λ is in $\boldsymbol{H}^\infty$ and, in particular, is in $\boldsymbol{H}^2$.

Now

$$
\begin{aligned}
(C_\phi f_\lambda)(z) &= \exp\left(i\frac{\theta_0}{\beta}\Gamma(\phi(z))\right) \\
&= \exp\left(i\frac{\theta_0}{\beta}F(\Gamma(z))\right) \qquad (\text{since } \Gamma\circ\phi = F\circ\Gamma) \\
&= \exp\left(i\frac{\theta_0}{\beta}\left(\Gamma(z)+\beta\right)\right) \qquad (\text{since } F(w) = w+\beta) \\
&= \exp\left(i\frac{\theta_0}{\beta}\Gamma(z)\right)\exp\left(i\frac{\theta_0}{\beta}\beta\right) \\
&= e^{i\theta_0}\exp\left(i\frac{\theta_0}{\beta}\Gamma(z)\right) \\
&= \lambda f_\lambda(z).
\end{aligned}
$$

This shows that every λ of modulus 1 is an eigenvalue of C_ϕ and the theorem is established. □

The composition operators induced by hyperbolic disk automorphisms are particularly interesting in several respects. In particular, it has been shown that the existence of nontrivial invariant subspaces for all restrictions of a hyperbolic composition operator would imply existence of invariant subspaces for all operators on Hilbert space; see [124].

We shall see that the study of hyperbolic disk automorphisms reduces to the study of those with fixed points 1 and -1. We therefore begin by considering such special hyperbolic disk automorphisms.

Lemma 5.4.7. *Let ϕ be a hyperbolic disk automorphism satisfying $\phi(1) = 1$ and $\phi(-1) = -1$. Then there exists a real number a between -1 and 1 such that*

$$\phi(z) = \frac{z-a}{1-az}$$

for all $z \in \mathbb{D}$. Moreover, $\phi'(1) = \frac{1+a}{1-a}$ and $\phi'(-1) = \frac{1-a}{1+a}$. One of the positive numbers $\phi'(1)$ and $\phi'(-1)$ is less than 1 and the other is greater than 1.

Proof. Since ϕ is a disk automorphism, there exists an a in $\mathbb{D}$ and a λ of modulus 1 such that

$$\phi(z) = \lambda\frac{a-z}{1-\overline{a}z}.$$

Now $\phi(1) = 1$ implies that $\lambda(a-1) = 1-\overline{a}$ and $\phi(-1) = -1$ implies that $\lambda(a+1) = -1-\overline{a}$. It follows that

$$\frac{a-1}{a+1} = \frac{\overline{a}-1}{\overline{a}+1}.$$

Thus $\frac{a-1}{a+1}$ is real and therefore a is real.

Now $\phi(1) = 1$ gives $\lambda(a-1) = 1-a$, so $\lambda = -1$. Therefore

$$\phi(z) = \frac{z-a}{1-az}$$

as required.

Differentiating gives

$$\phi'(z) = \frac{1-a^2}{(1-az)^2}.$$

Hence, $\phi'(1) = \frac{1+a}{1-a}$ and $\phi'(-1) = \frac{1-a}{1+a}$.

Since $a \in (-1,1)$, it is clear that both $\phi'(1)$ and $\phi'(-1)$ are positive. If either one of $\phi'(1)$ or $\phi'(-1)$ is 1, then $a = 0$ and $\phi(z) = z$ for all $z \in \mathbb{D}$. In the alternative, the fact that $\phi'(1)\phi'(-1) = 1$ implies that one factor is less than 1 and the other is greater than 1. □

As in the parabolic case, the iterates of a hyperbolic composition operator can be more easily studied by transforming the problem to the upper half-plane.

Lemma 5.4.8. *Let ϕ be a hyperbolic disk automorphism satisfying $\phi(1) = 1$ and $\phi(-1) = -1$. If $\Gamma(z) = i\frac{1+z}{1-z}$ and the function G is defined by $G = \Gamma \circ \phi \circ \Gamma^{-1}$, then there is a $\gamma > 0$ such that $G(z) = \gamma z$ for all z in the upper half-plane. Moreover, $\gamma = \phi'(-1)$.*

Proof. Note that G is an automorphism of the upper half-plane and that G maps the extended real line to the extended real line. Also

$$G(0) = \Gamma(\phi(-1)) = \Gamma(-1) = 0,$$

and

$$G(\infty) = \Gamma(\phi(1)) = \Gamma(1) = \infty.$$

Since G is a linear fractional transformation, it must have the form

$$G(z) = \frac{az+b}{cz+d}.$$

Since $G(\infty) = \infty$, we must have $c = 0$. Thus G has the form

$$G(z) = \gamma z + \delta,$$

for suitable γ and δ. However, $G(0) = 0$, so $\delta = 0$. Since G is an automorphism of the open upper half-plane, γ must be a positive real number.

We can compute γ as follows. Since $G \circ \Gamma = \Gamma \circ \phi$, we have

$$G'(\Gamma(-1))\, \Gamma'(-1) = \Gamma'(\phi(-1))\, \phi'(-1),$$

so

$$\gamma\, \Gamma'(-1) = \Gamma'(-1)\, \phi'(-1).$$

It is easily seen that $\Gamma'(-1)$ is not 0; hence $\gamma = \phi'(-1)$. □

Lemma 5.4.9. *Let $a \in (-1,1)$ and $\phi(z) = \dfrac{z-a}{1-az}$. Then*

$$r(C_\phi) = \sqrt{\phi'(1)} = \sqrt{\frac{1+a}{1-a}}$$

if $a \geq 0$ and

$$r(C_\phi) = \sqrt{\phi'(-1)} = \sqrt{\frac{1-a}{1+a}}$$

if $a \leq 0$.

Proof. Recall that, by Corollary 5.1.10,

$$\frac{1}{\sqrt{1-|\phi(0)|^2}} \leq \|C_\phi\| \leq \frac{2}{\sqrt{1-|\phi(0)|^2}}.$$

More generally, for every natural number n,

$$\left(\frac{1}{\sqrt{1-|\phi^{[n]}(0)|^2}}\right)^{1/n} \leq \|C_\phi^n\|^{1/n} \leq \left(\frac{2}{\sqrt{1-|\phi^{[n]}(0)|^2}}\right)^{1/n}.$$

Since $\lim_{n\to\infty} 2^{1/n} = 1$, it follows that

$$r(C_\phi) = \lim_{n\to\infty} \|C_\phi^n\|^{1/n} = \lim_{n\to\infty} \left(\frac{1}{\sqrt{1-|\phi^{[n]}(0)|^2}}\right)^{1/n}.$$

As in Lemma 5.4.8, define G to be $\Gamma \circ \phi \circ \Gamma^{-1}$. By Lemma 5.4.8, $G(z) = \gamma z$ for some $\gamma > 0$.

Clearly $G^{[n]}(z) = \gamma^n z$ for every natural number n. Since $\Gamma^{-1}(i) = 0$, we have

$$\phi^{[n]}(0) = \phi^{[n]}(\Gamma^{-1}(i)) = \Gamma^{-1}(G^{[n]}(i)) = \Gamma^{-1}(\gamma^n i) = \frac{\gamma^n i - i}{\gamma^n i + i} = \frac{\gamma^n - 1}{\gamma^n + 1}.$$

It follows that

$$1 - |\phi^{[n]}(0)|^2 = 1 - \left|\frac{\gamma^n - 1}{\gamma^n + 1}\right|^2 = \frac{(\gamma^n+1)^2 - (\gamma^n-1)^2}{(\gamma^n+1)^2} = \frac{4\gamma^n}{(\gamma^n+1)^2}.$$

Using the formula for the spectral radius obtained above, we get

$$\begin{aligned} r(C_\phi) &= \lim_{n\to\infty}\left(\frac{1}{\sqrt{1-|\phi^{[n]}(0)|^2}}\right)^{1/n} \\ &= \lim_{n\to\infty}\frac{(\gamma^n+1)^{1/n}}{2^{1/n}\sqrt{\gamma}} \\ &= \frac{1}{\sqrt{\gamma}}\lim_{n\to\infty}(\gamma^n+1)^{1/n}, \end{aligned}$$

since $\lim_{n\to\infty} 2^{1/n} = 1$.

We need to consider two cases: the case $0 < \gamma < 1$ and the case $\gamma > 1$. (Notice that $\gamma = 1$ corresponds to the case in which ϕ is the identity transformation.)

If $0 < \gamma < 1$, then $\lim_{n\to\infty}(\gamma^n + 1)^{1/n} = 1$, and thus

$$r(C_\phi) = \frac{1}{\sqrt{\gamma}}.$$

If $\gamma > 1$, then $\lim_{n\to\infty}(\gamma^n + 1)^{1/n} = \gamma$, and thus

$$r(C_\phi) = \sqrt{\gamma}.$$

By Lemma 5.4.8, $\gamma = \phi'(-1)$, and by Lemma 5.4.7, $\phi'(-1) = \frac{1-a}{1+a}$. Hence $\gamma = \frac{1-a}{1+a}$.

If $a \leq 0$, then $\gamma \geq 1$, and therefore

$$r(C_\phi) = \sqrt{\frac{1-a}{1+a}}.$$

If $a \geq 0$, then $\gamma \leq 1$, and therefore

$$r(C_\phi) = \sqrt{\frac{1+a}{1-a}}.$$

□

Theorem 5.4.10. *If ϕ is a hyperbolic disk automorphism, then ϕ has a unique fixed point α such that $0 < \phi'(\alpha) < 1$. The spectrum of C_ϕ is given by*

$$\sigma(C_\phi) = \left\{ z \,:\, \sqrt{\phi'(\alpha)} \leq |z| \leq \frac{1}{\sqrt{\phi'(\alpha)}} \right\}.$$

Moreover, every point of

$$\left\{ z \,:\, \sqrt{\phi'(\alpha)} < |z| < \frac{1}{\sqrt{\phi'(\alpha)}} \right\}$$

is an eigenvalue of C_ϕ.

Proof. We begin by reducing to the case that the fixed points are -1 and 1. Suppose that $\phi(\beta) = \beta$ and $\phi(\delta) = \delta$. Let T be a disk automorphism such that $T(1) = \beta$ and $T(-1) = \delta$. Define ψ by $\psi = T^{-1} \circ \phi \circ T$. Then ψ is a hyperbolic disk automorphism satisfying $\psi(1) = 1$ and $\psi(-1) = -1$. As in the case of elliptic automorphisms (see the proof of Theorem 5.4.3), C_ψ is similar to C_ϕ, so $\sigma(C_\psi) = \sigma(C_\phi)$ and $\Pi_0(C_\psi) = \Pi_0(C_\phi)$. Moreover, $\psi'(1) = \phi'(\beta)$ and $\psi'(-1) = \phi'(\delta)$. By Lemma 5.4.7, one of $\psi'(1)$ and $\psi'(-1)$ is between 0 and 1. Let α equal β or δ so that $\phi'(\alpha)$ is between 0 and 1.

Thus it suffices to prove the theorem for hyperbolic disk automorphisms ϕ whose fixed points are 1 and -1. Assume, then, that $\phi(1) = 1$, $\phi(-1) = -1$, and $\phi'(1)$ is between 0 and 1.

By Lemma 5.4.9, $r(C_\phi) = \sqrt{\phi'(-1)} = \frac{1}{\sqrt{\phi'(1)}}$. Hence

$$\sigma(C_\phi) \subset \left\{ z : |z| \leq \frac{1}{\sqrt{\phi'(1)}} \right\}.$$

Clearly, ϕ^{-1} is also a hyperbolic disk automorphism with fixed points 1 and -1. Moreover, $\left(\phi^{-1}\right)'(-1) = \frac{1}{\phi'(-1)}$, so Lemma 5.4.9 implies

$$r(C_{\phi^{-1}}) = \frac{1}{\sqrt{(\phi^{-1})'(-1)}} = \sqrt{\phi'(-1)} = \frac{1}{\sqrt{\phi'(1)}}.$$

Now, $C_{\phi^{-1}} = C_\phi^{-1}$ (by Theorem 5.2.1) and

$$\sigma(C_\phi^{-1}) = \left\{ \frac{1}{z} : z \in \sigma(C_\phi) \right\} \qquad \text{(by Theorem 1.2.4)}.$$

Therefore, $z \in \sigma(C_\phi)$ implies that

$$\frac{1}{|z|} \leq \frac{1}{\sqrt{\phi'(1)}}.$$

It follows that $|z| \geq \sqrt{\phi'(1)}$ for all z in $\sigma(C_\phi)$, so $\sigma(C_\phi)$ is contained in the specified annulus.

The proof of the theorem will be complete upon showing that every point in the open annulus

$$\left\{ z : \sqrt{\phi'(1)} < |z| < \frac{1}{\sqrt{\phi'(1)}} \right\}$$

is an eigenvalue of C_ϕ.

Fix any number $s \in \left(-\frac{1}{2}, \frac{1}{2}\right)$ and any real number t, and let

$$f(z) = \left(\frac{1+z}{1-z}\right)^{s+it}$$

(where the power is defined in terms of the principal branch of the logarithm). We will show that each such f is an eigenvector of C_ϕ and, moreover, that the corresponding eigenvalues range over all the complex numbers in the open annulus as s ranges over all real numbers in $(-\frac{1}{2}, \frac{1}{2})$ and t ranges over all real numbers.

It must first be established that each such function f is in $\boldsymbol{H}^2$. Note that

$$f(z) = \left(\frac{1+z}{1-z}\right)^{s} \left(\frac{1+z}{1-z}\right)^{it}.$$

Since

$$\left(\frac{1+z}{1-z}\right)^{it}$$

is in $\boldsymbol{H}^\infty$ (Example 1.1.18), it suffices to prove that the function

$$g(z) = \left(\frac{1+z}{1-z}\right)^{s}$$

is in $\boldsymbol{H}^2$ for every s in $(-\frac{1}{2}, \frac{1}{2})$.

For $s = 0$ there is nothing to prove. If s is in $(0, \frac{1}{2})$, then $\frac{1}{(1-z)^s}$ is in $\boldsymbol{H}^2$ (by Example 1.1.14) and $(1+z)^s$ is clearly in $\boldsymbol{H}^\infty$, so g is in $\boldsymbol{H}^2$.

For s in $(-\frac{1}{2}, 0)$,

$$g(z) = \left(\frac{1-z}{1+z}\right)^{-s}.$$

The function $(1-z)^{-s}$ is clearly in $\boldsymbol{H}^\infty$. The result of Example 1.1.14 implies that $\left(\frac{1}{1+z}\right)^{-s}$ is in $\boldsymbol{H}^2$ since $h(z)$ in $\boldsymbol{H}^2$ implies that $h(-z) \in \boldsymbol{H}^2$. Therefore g is in $\boldsymbol{H}^2$ for all s in $(-\frac{1}{2}, 0)$.

Thus

$$f(z) = \left(\frac{1+z}{1-z}\right)^{s+it}$$

is in $\boldsymbol{H}^2$ for all s in $(-\frac{1}{2}, \frac{1}{2})$ and all real numbers t.

We must compute $C_\phi f$ for such f.

Since ϕ has fixed points 1 and -1, there is a real number a between -1 and 1 such that

$$\phi(z) = \frac{z-a}{1-az} \qquad \text{(by Lemma 5.4.7).}$$

Using this form for ϕ yields

$$\frac{1+\phi(z)}{1-\phi(z)} = \frac{(1-a)+(1-a)z}{(1+a)-(1+a)z} = \frac{1-a}{1+a}\,\frac{1+z}{1-z}.$$

It follows that

$$(C_\phi f)(z) = (f\circ\phi)(z) = \left(\frac{1-a}{1+a}\,\frac{1+z}{1-z}\right)^{s+it} = \left(\frac{1-a}{1+a}\right)^{s+it} f(z).$$

Thus each such f is an eigenvector for C_ϕ and the eigenvalue that corresponds to such an f is

$$\left(\frac{1-a}{1+a}\right)^{s+it}.$$

Note that $\phi'(1) = \frac{1+a}{1-a}$ (by Lemma 5.4.7). All that remains to be shown is that the function λ defined by

$$\lambda(s,t) = \left(\frac{1}{\phi'(1)}\right)^{s+it}$$

assumes all values in the open annulus as s ranges over the interval $(-\frac{1}{2},\frac{1}{2})$ and t ranges over all real numbers. Note that

$$\lambda(s,t) = \left(\frac{1}{\phi'(1)}\right)^{s}\left(\frac{1}{\phi'(1)}\right)^{it}.$$

As t ranges over all real numbers, the function

$$\left(\frac{1}{\phi'(1)}\right)^{it}$$

ranges over all complex numbers of modulus 1. As s ranges over the interval $(-\frac{1}{2},\frac{1}{2})$, the function

$$\left(\frac{1}{\phi'(1)}\right)^{s}$$

ranges over the open interval

$$\left((\phi'(1))^{1/2}, \left(\frac{1}{\phi'(1)}\right)^{1/2}\right)$$

(recall that $\phi'(1) < 1$). Since the product of each of these eigenvalues with every complex number of modulus 1 is also an eigenvalue, it follows that every point in the open annulus is an eigenvalue of C_ϕ. □

5.5 Exercises

5.1. Show that $\phi(0) = 0$ implies that $z^n \boldsymbol{H}^2$ is invariant under C_ϕ for every natural number n.

5.2. Show that C_ϕ is a Hilbert–Schmidt operator whenever $\|\phi\|_\infty < 1$. (Recall that the operator A is *Hilbert–Schmidt* if there is an orthonormal basis $\{e_n\}$ such that $\sum_{n=0}^{\infty} \|Ae_n\|^2$ converges.)

5.3. Show that

$$\lim_{|z| \to 1^-} \frac{1 - |z|^2}{1 - |\phi(z)|^2} = 0$$

whenever C_ϕ is compact.

5.4. Let $\phi(z) = \frac{z+1}{2}$. Show that C_ϕ is not compact.

5.5. Show that an invertible composition operator maps inner functions to inner functions.

5.6. Let ϕ be an analytic function mapping $\mathbb{D}$ into itself and satisfying $\phi(a) = a$ for some $a \in \mathbb{D}$.

(i) First assume that $a = 0$. Show that $\{\phi^{[n]}\} \to a$ uniformly on compact subsets of $\mathbb{D}$ unless ϕ is of the form $\phi(z) = \lambda z$ for some λ of modulus 1. (Hint: Use the Schwarz lemma.)

(ii) Assume $a \neq 0$. Let ϕ_a be the linear fractional transformation defined by $\phi_a(z) = \frac{a-z}{1-\bar{a}z}$. Show that $\{\phi^{[n]}\} \to a$ uniformly on compact subsets of $\mathbb{D}$ unless ϕ is of the form $\phi(z) = \phi_a(\lambda \phi_a(z))$ for some λ of modulus 1.

5.7. Prove Lemma 5.4.8 by using the fact from Lemma 5.4.7 that ϕ has the form

$$\phi(z) = \frac{z - a}{1 - az}$$

for some real number a and then directly computing $\Gamma \circ \phi \circ \Gamma^{-1}$.

5.8. Let ϕ be a parabolic disk automorphism. Show that $\{\phi^{[n]}\}$ converges uniformly on compact subsets of $\mathbb{D}$ to the fixed point of ϕ. (Hint: See the proof of Theorem 5.4.5.)

5.9. Let ϕ be a hyperbolic disk automorphism. Show that $\{\phi^{[n]}\}$ converges uniformly on compact subsets of $\mathbb{D}$ to one of the fixed points of ϕ and that the iterates of ϕ^{-1} converge uniformly on compact subsets of $\mathbb{D}$ to the other fixed point. (Hint: See the proofs of Lemma 5.4.8 and Theorem 5.4.10.)

5.10. Show that C_ϕ is similar to λC_ϕ whenever ϕ is a hyperbolic disk automorphism and λ is a complex number of modulus 1.

5.11. If ϕ is an inner function show that

$$\|C_\phi\| = \sqrt{\frac{1+|\phi(0)|}{1-|\phi(0)|}}.$$

5.12. Prove that the operator C_ϕ is an isometry if and only if ϕ is inner and $\phi(0) = 0$.

5.13. An operator A is said to be *quasinormal* if A commutes with A^*A. Show that $\phi(0) = 0$ if the composition operator C_ϕ is quasinormal.

5.14. Let $\phi(z) = \frac{az+b}{cz+d}$ be any nonconstant linear fractional transformation mapping $\mathbb{D}$ into itself.

(i) Show that ϕ nonconstant implies that $ad - bc \neq 0$. Conclude that any nonconstant such ϕ can be represented in the given form with $ad - bc = 1$.

For the rest of this problem, assume that $ad - bc = 1$.

(ii) Prove that the linear fractional transformation ψ defined by $\psi(z) = \frac{\overline{a}z-\overline{c}}{-\overline{b}z+\overline{d}}$ maps $\mathbb{D}$ into itself.

(iii) Define g and h by $g(z) = \frac{1}{-\overline{b}z+\overline{d}}$ and $h(z) = cz + d$. Show that g and h are in $\boldsymbol{H}^\infty$.

(iv) Prove that $C_\phi^* = T_g C_\psi T_h^*$.

5.15. Let T_ψ be a nonzero analytic Toeplitz operator and C_ϕ be a composition operator. Prove that $\psi \circ \phi = \psi$ if and only if C_ϕ and T_ψ commute.

5.16. The Berezin symbol of an operator on $\boldsymbol{H}^2$ was defined in Exercise 1.17 in Chapter 1.

(i) Compute the Berezin symbol of C_ϕ in the case $\phi(z) = z\psi(z)$ for ψ an analytic function mapping the disk into itself.

(ii) Let $\mathcal{R}$ be any open, connected, and simply connected subset of $\mathbb{D}$ whose boundary is a simple closed Jordan curve that does not contain 1 and whose intersection with S^1 contains a nontrivial arc. Note that the Riemann mapping theorem implies that there are conformal mappings from

$\mathbb{D}$ onto $\mathcal{R}$. Let ψ be such a conformal mapping and let ϕ be defined by $\phi(z) = z\psi(z)$. Show that

$$\lim_{|z|\to 1^-} \widetilde{C_\phi}(z) = 0$$

but that C_ϕ is not compact.

5.6 Notes and Remarks

Composition of functions is one of the most common and fundamental operations in mathematics. Thus many aspects of composition operators implicitly date back to the beginnings of complex analysis. More explicitly, in the late nineteenth century, Schröder [154] and Königs [108] studied the solutions of the equation that, in particular, defines the eigenvectors of composition operators (see Theorem 5.3.3). Littlewood's subordination theorem goes back to 1925 (see [111]). His theorem is much more general than the special case that we have presented in Corollary 5.1.8; in particular, it applies to functions in $\boldsymbol{H}^p$. The upper bound on the norm of composition operators given in Theorem 5.1.5 was first established in Ryff [147], which also contains some interesting related results.

The idea of studying the general properties of composition operators is due to Eric Nordgren [121]. In particular, Nordgren [121] determined the spectra of composition operators induced by disk automorphisms (i.e., Theorems 5.4.3, 5.4.6, and 5.4.10).

Nordgren suggested the study of composition operators to Peter Rosenthal, who, in turn, passed on that suggestion to his Ph.D. student Howard Schwartz. Schwartz's thesis [155] contained a number of fundamental results, including theorems on compactness of composition operators (Theorem 5.1.16, Theorem 5.1.17, and Exercise 5.4), the lower bound on the norm of C_ϕ given in Theorem 5.1.10, and results on spectra such as Theorem 5.3.1. Moreover, Schwartz obtained Theorem 5.1.13, Corollary 5.1.14, and Theorem 5.1.15. Although Schwartz's thesis was never published, it was widely circulated and greatly stimulated interest in composition operators.

Exercise 5.3 is an unpublished result of Nordgren's; Theorem 5.3.5 is due to Caughran and Schwartz [69]; and Exercise 5.10 is taken from Cowen and MacCluer [14]. Exercise 5.11 is due to Nordgren [121]. Exercise 5.12 was established under the assumption that $\phi(0) = 0$ by Ryff [147]; the general case is due to Schwartz [155]. Exercise 5.13 is a result of Cload [73], while Exercise 5.14 was shown by Cowen [77]. Exercise 5.16 is due to Nordgren and Rosenthal [122].

We have presented only a fragment of what is known about composition operators; there are many other interesting results. In particular, Joel Shapiro [157] obtained a necessary and sufficient condition that C_ϕ be compact in terms of the Nevanlinna counting function of ϕ. Shapiro's book [53] contains an exposition of this and a number of other results. The book by Carl

Cowen and Barbara MacCluer [14] contains numerous results about spectra of composition operators as well as many other theorems concerning composition operators on $\boldsymbol{H}^2$ and on other spaces of analytic functions.

Chapter 6

Further Reading

This book is merely an introduction to a vast subject. There is a great deal of additional knowledge concerning these topics, and there are a number of excellent books and expository papers treating much of that material. We briefly describe some of these expositions in order to make it easier for the reader to pursue further study.

The space $\boldsymbol{H}^2$ is one of the $\boldsymbol{H}^p$ spaces. The reader interested in learning more about $\boldsymbol{H}^2$ and in learning about all the $\boldsymbol{H}^p$ spaces might begin by studying Duren [17] or Hoffman [32]. Books containing additional results include those of Garnett [22], Koosis [33], and Nikolskii [35].

There are other interesting linear spaces of analytic functions. In particular, there has been a lot of research concerning Bergman spaces and operators on them: see Duren–Schuster [18] and Hedenmalm–Korenblum–Zhu [31]. These and other spaces of analytic functions are discussed in Zhu [57] and Cowen–MacCluer [14]. Linear spaces of analytic functions of several variables have also been studied; see, for example, Rudin [45]. There are analogous linear spaces whose elements are harmonic functions instead of analytic functions; see Duren [17], Garnett [22], and, especially, Axler–Bourdon–Ramey [5].

There is an interesting Banach space called BMO, the space of functions of bounded mean oscillation, that is closely related to the Hardy spaces $\boldsymbol{H}^1$ and $\boldsymbol{H}^\infty$. Certain properties of Hankel operators can be characterized in terms of functions in BMO. Nice introductions to BMO are contained in Chapter 6 of Garnett [22] and in Chapter X of Koosis [33]. Theorem 1.2 in [38] characterizes the boundedness of a Hankel operator in terms of BMO; for other connections between BMO and Hankel operators, see Nikolskii [35].

There are many fine books concerning linear operators on Hilbert space. Two that are particularly readable are Halmos [27] and Conway [13]. For more information about invariant subspaces, see Radjavi–Rosenthal [41].

The unilateral shift, and unilateral shifts of higher multiplicity (i.e., direct sums of copies of the unilateral shift), have many interesting properties and have been much studied. Three of the many books that contain additional results are Helson [29], Nikolskii [34], and Cima–Ross [8]. Clear treatments of the related subject of interpolation problems can be found in Rosenblum–Rovnyak [43] and Agler-McCarthy [1] (also see Nikolskii [35]). The unilateral shift is a particular example of a "weighted unilateral shift". A beautiful discussion of weighted shifts and their relationship to weighted Hardy–Hilbert spaces is given by Allen Shields in [158].

Don Sarason [153] has written an extremely nice expository paper concerning operators on spaces of analytic functions. Although this paper does not contain proofs, it provides a remarkably instructive overview of the subject and contains an excellent bibliography.

Ron Douglas [16] gives a very clear exposition of many results about Toeplitz operators, including the deep theorem of Widom [170] that the spectrum of every Toeplitz operator is connected. (Douglas's proof of that result has subsequently been simplified in Searcóid [156].) A number of additional theorems concerning Toeplitz operators can be found in Nikolskii [35]. Sarason has written a set of lecture notes [49] that contains a very nice account of much of the theory of Toeplitz operators. Unfortunately, these lecture notes are not readily available; they make very valuable reading if they can be obtained.

A reader interested in learning more about Hankel operators could begin by studying Steve Power's expository article [141] and text [40], both of which provide very clear treatments of the subject. This could be followed by reading the books of Peller [38] and Nikolskii [35].

There are many other interesting theorems about composition operators. Joel Shapiro has written a very nice book [53] that contains a thorough discussion of compactness of composition operators, as well as other topics, including hypercyclicity. Shapiro's book is written in a very clear and accessible manner. The book [14] by Carl Cowen and Barbara MacCluer is an excellent, encyclopedic, treatment of composition operators. It contains a wide variety of results about composition operators on a wide variety of spaces of analytic functions. There are numerous theorems about boundedness and about spectral properties of composition operators on various spaces. Moreover, [14] has an extensive bibliography.

Control theory is an important part of engineering. One approach to control theory is through "$\boldsymbol{H}^\infty$ methods"; this approach relies on some of the material we have discussed, including properties of Hankel operators. A brief introduction to this subject can be found in Partington [36, Chapter 5]. More extensive treatments are contained in Francis [21], Helton–Merino [30], and Sasane [52].

In addition to the expository writing referred to above, there are many research papers on these topics, some of which are mentioned in the "Notes and Remarks" sections of the previous chapters of this book. Additional references are listed in the bibliography that follows this chapter. It is suggested that readers continue their study by perusing those references and, as Halmos [96] says, "iterating the bibliography operator".

References

Books

1. Jim Agler and John E. McCarthy, *Pick Interpolation and Hilbert Function Spaces*, American Mathematical Society, Providence, 2002.
2. N. I. Akhiezer and I. M. Glazman, *Theory of Linear Operators in Hilbert Space*, Dover, Mineola, NY, 1993.
3. Tom M. Apostol, *Mathematical Analysis*, second edition, Addison Wesley, Reading, MA, 1974.
4. Tom M. Apostol, *Introduction to Analytic Number Theory*, Springer, New York, 1976.
5. Sheldon Axler, Paul Bourdon, and Wade Ramey, *Harmonic Function Theory*, second edition, Springer, New York, 2001.
6. James W. Brown and Ruel V. Churchill, *Complex Variables and Applications*, Seventh Edition, McGraw Hill, New York, 2003.
7. Lennart Carleson and Theodore W. Gamelin, *Complex Dynamics*, Springer, New York, 1993.
8. Joseph Cima and William Ross, *The Backward Shift on the Hardy Space*, American Mathematical Society, Providence, 2000.
9. John B. Conway, *Functions of One Complex Variable*, second edition, Springer, New York, 1978.
10. John B. Conway, *Subnormal Operators*, Pittman, London, 1981.
11. John B. Conway, *The Theory of Subnormal Operators*, American Mathematical Society, Providence, 1991.
12. John B. Conway, *A Course in Functional Analysis*, second edition, Springer, New York, 1997.
13. John B. Conway, *A Course in Operator Theory*, American Mathematical Society, Providence, 2000.

14. Carl Cowen and Barbara MacCluer, *Composition Operators on Spaces of Analytic Functions*, CRC Press, Boca Raton, FL, 1995.
15. Kenneth Davidson, *Nest Algebras*, Longman, copublished with John Wiley, New York, 1988.
16. Ronald G. Douglas, *Banach Algebra Techniques in Operator Theory*, second edition, Springer, New York, 1998.
17. Peter L. Duren, *Theory of H^p Spaces*, Dover, Mineola, NY, 2000.
18. Peter Duren and Alexander Schuster, *Bergman Spaces*, American Mathematical Society, Providence, 2004.
19. Nelson Dunford and Jacob T. Schwartz, *Linear Operators. Part I: General Theory*, Interscience, New York, 1967.
20. Gerald B. Folland, *Real Analysis*, Wiley, New York, 1984.
21. Bruce A. Francis, *A Course in H_∞ Control Theory*, Springer, Berlin, 1987.
22. John B. Garnett, *Bounded Analytic Functions*, Academic Press, New York, 1981.
23. G. M. Goluzin, *Geometric Theory of Functions of a Complex Variable*, Translations of Mathematical Monographs, AMS, Providence, 1969.
24. G. H. Hardy, *A Mathematician's Apology*, Cambridge University Press, London, 1967.
25. G. H. Hardy, J. E. Littlewood, and G. Pólya, *Inequalities*, reprint of the 1952 edition, Cambridge University Press, Cambridge, 1988.
26. G. H. Hardy and E. M. Wright, *An Introduction to the Theory of Numbers*, fifth edition, the Clarendon Press, Oxford University Press, New York, 1979.
27. Paul R. Halmos, *A Hilbert Space Problem Book*, second edition, Springer, New York, 1982.
28. Paul R. Halmos, *Introduction to Hilbert Space*, reprint of the second edition (1957), AMS Chelsea Publishing, Providence, 1998.
29. Henry Helson, *Lectures on Invariant Subspaces*, Academic Press, New York, 1964.
30. J. William Helton and Orlando Merino, *Classical Control Using $\boldsymbol{H}^\infty$ Methods: Theory, Optimization and Design*, SIAM, Philadelphia, 1998.
31. Haakan Hedenmalm, Boris Korenblum, and Kehe Zhu, *Theory of Bergman Spaces*, Springer, New York, 2000.
32. Kenneth Hoffman, *Banach Spaces of Analytic Functions*, Dover, New York, 1988.
33. Paul Koosis, *Introduction to H_p Spaces*, second edition, Cambridge University Press, Cambridge, 1998.
34. Nikolai K. Nikolskii, *Treatise on the Shift Operator*, Springer, Berlin, 1986.
35. Nikolai K. Nikolskii, *Operators, Functions and Systems: An Easy Reading. Volume 1: Hardy, Hankel, and Toeplitz*, American Mathematical Society, Providence, 2002.

36. Jonathan R. Partington, *An Introduction to Hankel Operators*, Cambridge University Press, Cambridge, 1988.
37. Pasquale Porcelli, *Linear Spaces of Analytic Functions*, Rand McNally, Chicago, 1966.
38. Vladimir V. Peller, *Hankel Operators and Their Applications*, Springer, New York, 2003.
39. I. I. Privalov, *Randeigenschaften analytischer Funktionen*, Deutscher Verlag der Wissenschaften, 1956.
40. Stephen Power, *Hankel Operators on Hilbert Space*, Pitman, Boston, 1982.
41. Heydar Radjavi and Peter Rosenthal, *Invariant Subspaces*, second edition, Dover, Mineola, NY, 2003.
42. Michael Reed and Barry Simon, *Methods of Modern Mathematical Physics. I: Functional Analysis*, revised and enlarged edition, Academic Press, New York, 1980.
43. Marvin Rosenblum and James Rovnyak, *Hardy Classes and Operator Theory*, Corrected reprint of the 1985 original, Dover, Mineola, NY, 1997.
44. H. L. Royden, *Real Analysis*, third edition, Prentice Hall, Englewood Cliffs, NJ, 1988.
45. Walter Rudin, *Function Theory in Polydiscs*, W. A. Benjamin, Inc., New York–Amsterdam, 1969.
46. Walter Rudin, *Principles of Mathematical Analysis*, third edition, McGraw Hill, New York, 1976.
47. Walter Rudin, *Real and Complex Analysis*, third edition, McGraw Hill, New York, 1986.
48. Walter Rudin, *Functional Analysis*, second edition, McGraw Hill, New York, 1991.
49. Donald Sarason, *Function Theory on the Unit Circle*, Notes for Lectures given at a Conference at Virginia Polytechnic Institute and State University, Blacksburg, Va., June 19–23, 1978. Virginia Polytechnic Institute and State University, Department of Mathematics, Blacksburg, Va., 1978.
50. Donald Sarason, *Sub-Hardy Hilbert Spaces*, John Wiley & Sons, New York, 1994.
51. Donald Sarason, *Notes on Complex Function Theory*, Helson, 1994.
52. Amol J. Sasane, *Hankel Norm Approximation for Infinite-Dimensional Systems*, Lecture Notes in Control and Information Sciences, 277, Springer, Berlin, 2002.
53. Joel H. Shapiro, *Composition Operators and Classical Function Theory*, Springer, New York, 1993.
54. Béla Sz.-Nagy and Ciprian Foiaş, *Harmonic Analysis of Operators on Hilbert Space*, North-Holland, Amsterdam, 1970.
55. Angus E. Taylor and David C. Lay, *Introduction to Functional Analysis*, second edition, John Wiley, New York, 1980.

56. Takahashi Yoshino, *Algebraic Properties and Relations of Twin Toeplitz and Hankel Operators*, Mathematical Institute, Tôhoku University, 2001.
57. Kehe Zhu, *Operator Theory in Function Spaces*, Marcel Dekker, 1990.

Articles

58. V. M. Adamjan, D. Z. Arov, and M. G. Kreĭn, *Infinite Hankel block matrices and related problems of extension*, Izv. Akad. Nauk Armjan. SSR Ser. Mat. **6** (1971) 87–112.
59. Alexei B. Aleksandrov and Vladimir V. Peller, *Hankel operators and similarity to a contraction*, Internat. Math. Res. Notices (1996) no. 6, 263–275.
60. N. Aronszajn and K. T. Smith, *Invariant subspaces of completely continuous operators*, Ann. of Math. (2) **60** (1954) 345–350.
61. Sheldon Axler, Sun-Yun A. Chang and Donald Sarason, *Products of Toeplitz operators*, Integral Equations Operator Theory **1** (1978) 285–309.
62. José Barría and Paul R. Halmos, *Asymptotic Toeplitz operators*, Trans. Amer. Math. Soc. **273** (1982) 621–630.
63. Richard Bellman, *A note on the divergence of series*, Amer. Math. Monthly **50** (1943) 318–319.
64. Arne Beurling, *On two problems concerning linear transformations in Hilbert space*, Acta Math. **81** (1949) 239–255.
65. Wilhelm Blaschke, *Eine Erweiterung des Satzes von Vitali über Folgen analytischer Funktionen*, S.-B. Sächs. Akad. Wiss. Leipzig Math.-Natur. Kl. **67** (1915) 194–200.
66. F. F. Bonsall and Stephen Power, *A proof of Hartman's theorem on compact Hankel operators*, Math. Proc. Cambridge Philos. Soc. **78** (1975) 447–450.
67. Arlen Brown and Paul R. Halmos, *Algebraic properties of Toeplitz operators*, J. Reine Angew. Math. **213** (1963/1964) 89–102.
68. Alberto Calderón, Frank Spitzer, and Harold Widom, *Inversion of Toeplitz matrices*, Illinois J. Math. **3** (1959) 490–498.
69. James G. Caughran and Howard J. Schwartz, *Spectra of compact composition operators*, Proc. Amer. Math. Soc. **51** (1975) 127–130.
70. Man-Duen Choi, *Tricks or treats with the Hilbert matrix*, Amer. Math. Monthly **90** (1983) 301–312.
71. James A. Clarkson, *On the series of prime reciprocals*, Proc. Amer. Math. Soc. **17** (1966) 541.
72. Bruce Cload, *Generating the commutant of a composition operator* in Studies on Composition Operators (Laramie, WY, 1996), 11–15, Contemp. Math., 213, Amer. Math. Soc., Providence, RI, 1998.
73. Bruce Cload, *Composition operators: hyperinvariant subspaces, quasi-normals and isometries*, Proc. Amer. Math. Soc. **127** (1999) 1697–1703.

74. Bruce Cload, *Toeplitz operators in the commutant of a composition operator*, Studia Math. **133** (1999) 187–196.
75. Lewis A. Coburn, *Weyl's theorem for nonnormal operators*, Michigan Math. J. **13** (1966) 285–288.
76. Carl C. Cowen, *The commutant of an analytic Toeplitz operator*, Trans. Amer. Math. Soc. **239** (1978) 1–31.
77. Carl C. Cowen, *Linear fractional composition operators on $\boldsymbol{H}^2$*, Integral Equations Operator Theory **11** (1988) 151–160.
78. Carl C. Cowen and Barbara MacCluer, *Schroeder's equation in several variables*, Taiwanese J. Math. **7** (2003) 129–154.
79. Kenneth R. Davidson and Vern I. Paulsen, *Polynomially bounded operators*, J. Reine Angew. Math. **487** (1997) 153–170.
80. Chandler Davis, *The Toeplitz–Hausdorff theorem explained*, Canad. Math. Bull. **14** (1971) 245–246.
81. Chandler Davis, *An extremal problem for extensions of a sesquilinear form*, Linear Algebra and Appl. **13** (1976) 91–102.
82. Chandler Davis, William M. Kahan, and Hans F. Weinberger, *Norm-preserving dilations and their applications to optimal error bounds*, SIAM J. Numer. Anal. **19** (1982) 445–469.
83. Allen Devinatz, *Toeplitz operators on H^2 spaces*, Trans. Amer. Math. Soc. **112** (1964) 304–317.
84. Jacques Dixmier, *Les opérateurs permutables á l'opérateur integral*, Fas. 2. Portugal. Math. **8** (1979) 73–84.
85. Ronald G. Douglas, *On majorization, factorization, and range inclusion of operators on Hilbert space*, Proc. Amer. Math. Soc. **17** (1966) 413–415.
86. Ronald G. Douglas, *Banach algebra techniques in the theory of Toeplitz operators*, Expository Lectures from the CBMS Regional Conference held at the University of Georgia, Athens, Ga., June 12–16, 1972. CBMS Regional Conference Series in Mathematics, AMS, Providence, R.I., 1973.
87. Ronald G. Douglas, Harold S. Shapiro, and Allen L. Shields, *Cyclic vectors and invariant subspaces for the backward shift operator*, Ann. Inst. Fourier (Grenoble) **20** (1970) 37–76.
88. Konstantin M. Dyakonov, *Kernels of Toeplitz operators via Bourgain's factorization theorem*, J. Funct. Anal. **170** (2000) 93–106.
89. Per Enflo, *On the invariant subspace problem for Banach spaces*, Acta Math. **158** (1987) 213–313.
90. Leonhard Euler, *Variae observationes circa series infinitas*, Commentarii Academiae Scientiarum Imperialis Petropolitanae **9** (1737) 160–188.
91. Tamás Erdélyi and William Johnson, *The "full Müntz theorem" in $L_p[0,1]$ for $0 < p < \infty$*, J. Anal. Math. **84** (2001) 145–172.
92. P. Fatou, *Séries trigonométriques et séries de Taylor*, Acta Math. **30** (1906) 335–400.

93. Otto Frostman, *Potential de'equilibre et capacité des ensembles avec quelques applications à la théorie des fonctions*, Medd. Lunds Univ. Mat. Sem. **3** (1935) 1–118.
94. I. C. Gohberg and N. Ja. Krupnik, *The algebra generated by the Toeplitz matrices* Funkcional. Anal. i Priložen **3** (1969) 46–56.
95. Paul R. Halmos, *Normal dilations and extensions of operators*, Summa Brasil. Math. **2** (1950) 125–134.
96. Paul R. Halmos, *Ten years in Hilbert space*, Integral Equations Operator Theory **2** (1979) 529–564.
97. Hermann Hankel, *Über eine besondere Classe der symmetrischen Determinanten*, dissertation, Göttingen, 1861.
98. G. H. Hardy, *The mean value theorem of the modulus of an analytic function*, Proc. London Math. Soc. **14** (1915) 269–277.
99. Philip Hartman, *On completely continuous Hankel matrices*, Proc. Amer. Math. Soc. **9** (1958) 862–866.
100. Philip Hartman and Aurel Wintner, *On the spectra of Toeplitz's matrices*, Amer. J. Math. **72** (1950) 359–366.
101. Philip Hartman and Aurel Wintner, *The spectra of Toeplitz's matrices*, Amer. J. Math. **76** (1954) 867–882.
102. Andreas Hartmann, Donald Sarason, and Kristian Seip, *Surjective Toeplitz operators*, Acta Sci. Math. (Szeged) **70** (2004) 609–621.
103. Eric Hayashi, *The kernel of a Toeplitz operator*, Integral Equations Operator Theory **9** (1986) 588–591.
104. Eric Hayashi, *Classification of nearly invariant subspaces of the backward shift*, Proc. Amer. Math. Soc. **110** (1990) 441–448.
105. David Hilbert, *Grundzüge einer allgemeinen Theorie der linearen Integralgleichungen IV*, Nachr. Akad. Wiss. Göttingen. Math.–Phys. Kl. (1906) 157–227.
106. Gustav Herglotz, *Über Potenzreihen mit positivem, reellen Teil im Einheitskreis*, S.-B. Sächs. Akad. Wiss. Leipzig Math.–Natur. Kl. **63** (1911) 501–511.
107. Gaston Julia, *Sur la représentation analytique des opérateurs bornés ou fermés de l'espace hilbertien*, C. R. Acad. Sci. Paris **219** (1944) 225–227.
108. G. Königs, *Recherches sur les intégrales de certaines équations fonctionnelles*, Annales Scientifiques de l'École Normale Supérieure Sér. 3, **1** (1884) 3–41 (supplement).
109. M. G. Kreĭn, *Integral equations on the half-line with a kernel depending on the difference of the arguments*, Uspehi Mat. Nauk **13** (1958) 3–120.
110. Leopold Kronecker, *Zur Theorie der Elimination einer Variablen aus zwei algebraischen Gleichungen*, in Leopold Kronecker's Werke, Chelsea, New York, 1968.
111. J. E. Littlewood, *On inequalities in the theory of functions*, Proc. London Math. Soc. **23** (1925) 481–519.

112. Ali Mahvidi, *Invariant subspaces of composition operators*, J. Operator Theory **46** (2001) 453–476.
113. Rubén A. Martínez-Avendaño, *When do Toeplitz and Hankel operators commute?* Integral Equations Operator Theory **37** (2000) 341–349.
114. Rubén A. Martínez-Avendaño, *A generalization of Hankel operators*, J. Funct. Anal. **190** (2002) 418–446.
115. Rubén A. Martínez-Avendaño, *Essentially Hankel operators*, J. London Math. Soc. (2) **66** (2002) 741–752.
116. Rubén A. Martínez-Avendaño and Sergei Treil, *An inverse spectral problem for Hankel operators*, J. Operator Theory **48** (2002) 83–93.
117. Rubén A. Martínez-Avendaño and Peter Yuditskii, *Non-compact λ-Hankel operators*, Z. Anal. Anwendungen **21** (2002) 891–899.
118. Alexandre V. Megretskiĭ, Vladimir V. Peller, and Sergei R. Treil, *The inverse spectral problem for self-adjoint Hankel operators*, Acta Math. **174** (1995) 241–309.
119. C. Müntz, *Über den Approximationsatz von Weierstrass*, H.A. Schwartz Festschrift, Berlin (1914).
120. Zeev Nehari, *On bounded bilinear forms*, Ann. of Math. (2) **65** (1957) 153–162.
121. Eric Nordgren *Composition operators*, Canadian J. Math. **20** (1968) 442–449.
122. Eric Nordgren and Peter Rosenthal, *Boundary values of Berezin symbols*, in Nonselfadjoint operators and related topics (Beer Sheva, 1992), 362–368, Oper. Theory Adv. Appl., 73, Birkhäuser, Basel, 1994.
123. Eric Nordgren, Peter Rosenthal, and F. S. Wintrobe, *Composition operators and the invariant subspace problem*, C. R. Math. Rep. Acad. Sci. Canada **6** (1984) 279–283.
124. Eric Nordgren, Peter Rosenthal, and F. S. Wintrobe, *Invertible composition operators on H^p*, J. Funct. Anal. **73** (1987) 324–344.
125. Stephen Parrott, *On a quotient norm and the Sz.-Nagy–Foiaş lifting theorem*, J. Funct. Anal. **30** (1978) 311–328.
126. Vladimir V. Peller, *Hankel operators of class $\mathcal{S}_p$ and their applications (rational approximation, Gaussian processes, the problem of majorization of operators)*, Mat. Sb. (N.S.) **113(155)** (1980) 538–581.
127. Vladimir V. Peller, *Smooth Hankel operators and their applications (ideals $\mathcal{S}_p$, Besov classes, random processes)* Dokl. Akad. Nauk SSSR **252** (1980) 43–48.
128. Vladimir V. Peller, *Vectorial Hankel operators, commutators and related operators of the Schatten-von Neumann class γ_p*, Integral Equations Operator Theory **5** (1982) 244–272.
129. Vladimir V. Peller, *Description of Hankel operators on the class $\mathcal{S}_p$ for $p > 0$, investigation of the rate of rational approximation and other applications*, Math. Sb. (N.S.) **122(164)** (1983) 481–510.

130. Vladimir V. Peller, *Invariant subspaces for Toeplitz operators*, Zap. Nauchn. Sem. Leningrad. Otdel. Mat. Inst. Steklov. (LOMI) **126** (1983) 170–179.
131. Vladimir V. Peller, *Spectrum, similarity, and invariant subspaces of Toeplitz operators*, Izv. Akad. Nauk SSSR Ser. Mat. **50** (1986) 776–787.
132. Vladimir V. Peller, *Invariant subspaces of Toeplitz operators with piecewise continuous symbols*, Proc. Amer. Math. Soc. **119** (1993) 171–178.
133. Vladimir V. Peller, *An excursion into the theory of Hankel operators*, in Holomorphic spaces (Berkeley, CA, 1995), pp. 65–120, S. Axler, J. McCarthy, and D. Sarason (eds.), Math. Sci. Res. Inst. Publ., 33, Cambridge University Press, Cambridge, 1998.
134. Vladimir V. Peller and Sergei V. Khrushchëv, *Hankel operators, best approximations and stationary Gaussian processes*, Uspekhi Mat. Nauk **37** (1982) 53–124.
135. Gilles Pisier, *A polynomially bounded operator on Hilbert space which is not similar to a contraction*, J. Amer. Math. Soc. **10** (1997) 351–369.
136. Stephen Power, *Hankel operators with discontinuous symbol*, Proc. Amer. Math. Soc. **65** (1977) 77–79.
137. Stephen Power, *The essential spectrum of Hankel operators with piecewise continuous symbols*, Mich. Math. J. **25** (1978) 117–121.
138. Stephen Power, *Hankel operators with PQC symbols and singular integral operators*, Integral Equations Operator Theory **1** (1978) 448–452.
139. Stephen Power, *C^*-algebras generated by Hankel operators and Toeplitz operators*, J. Funct. Anal. **31** (1979) 52–68.
140. Stephen Power, *Hankel operators with PQC symbols and singular integral operators*, Proc. London Math. Soc. (3) **41** (1980) 45–65.
141. Stephen Power, *Hankel operators on Hilbert space*, Bull. London Math. Soc. **12** (1980) 422–442.
142. C. J. Read, *A solution to the invariant subspace problem on the space l_1*, Bull. London Math. Soc. **17** (1985) 305–317.
143. C. J. Read, *A short proof concerning the invariant subspace problem*, J. London Math. Soc. (2) **34** (1986) 335–348.
144. F. Riesz, *Über die Randwerte einer analytischen Funktion*, Math. Z. **18** (1923) 87–95.
145. F. Riesz and M. Riesz, *Über die Randwerte einer analytischen Funktion*, Quatrième Congrès de Math. Scand. Stockholm (1916) 27–44.
146. Peter Rosenthal, *Examples of invariant subspace lattices*, Duke Math. J. **37** (1970) 103–112.
147. John V. Ryff, *Subordinate H^p functions*, Duke Math. J. **33** (1966) 347–354.
148. Donald Sarason, *A remark on the Volterra operator*, J. Math. Anal. Appl. **12** (1965) 244–246.

149. Donald Sarason, *Generalized interpolation in H^∞*, Trans. Amer. Math. Soc. **127** (1967) 179–203.
150. Donald Sarason, *Kernels of Toeplitz operators*, in Toeplitz operators and related topics (Santa Cruz, CA, 1992), 153–164, Oper. Theory Adv. Appl., 71, Birkhäuser, Basel, 1994.
151. Donald Sarason, *Shift-invariant spaces from the Brangesian point of view*, in The Bieberbach conjecture (West Lafayette, Ind., 1985), 153–166, Math. Surveys Monogr., 21, AMS, Providence, RI, 1986.
152. Donald Sarason, *Angular derivatives via Hilbert space*, Complex Variables Theory Appl. **10** (1988) 1–10.
153. Donald Sarason, *Holomorphic spaces: a brief and selective survey*, in Holomorphic spaces (Berkeley, CA, 1995), 1–34, Math. Sci. Res. Inst. Publ., 33, Cambridge Univ. Press, Cambridge, 1998.
154. E. Schröder, *Über iterierte Funktionen*, Math. Ann. **3** (1871) 296–322.
155. Howard J. Schwartz, *Composition operators on H^p*, dissertation, University of Toledo, 1969.
156. Mícheál Ó Searcóid, *On a comment of Douglas concerning Widom's theorem*, Irish Math. Soc. Bull. **40** (1998) 31–34.
157. Joel H. Shapiro, *The essential norm of a composition operator*, Annals of Math. **125** (1987) 375–404.
158. Allen L. Shields, *Weighted shift operators and analytic function theory*, in Topics in operator theory, pp. 49–128. Math. Surveys, No. 13, Amer. Math. Soc., Providence, R.I., 1974.
159. V. I. Smirnov, *Sur les valeurs limites des fonctions, régulières à l'intérieur de'un cercle*, Journal de la Societé Phys.-Math. de Léningrade **2** (1929) 22–37.
160. Béla Sz.-Nagy, *Sur les contractions de l'espace de Hilbert*, Acta Sci. Math. Szeged **15** (1953) 87–92.
161. Otto Szász, *Über die Approximation stetiger Funktionen durch lineare Aggregate von Potenzen*, Math. Ann. **77** (1916) 482–496.
162. Otto Toeplitz, *Zur theorie der quadratischen Formen von unendlichvielen Veränderlichen*, Math. Ann. **70** (1911) 351–376.
163. Sergei R. Treil, *Moduli of Hankel operators and the V. V. Peller–S. Kh. Khrushchev problem*, Zap. Nauchn. Sem. Leningrad. Otdel. Mat. Inst. Steklov. (LOMI) **141** (1985) 188–189.
164. Sergei R. Treil, *Moduli of Hankel operators and the V. V. Peller–S. V. Khrushchev problem*, Dokl. Akad. Nauk SSSR **283** (1985) 1095–1099.
165. Sergei R. Treil, *An inverse spectral problem for the modulus of the Hankel operator, and balanced realizations*, Algebra i Analiz **2** (1990) 158–182; translation in Leningrad Math. J. **2** (1991) 353–375.
166. Sergie R. Treil and Vasily I. Vasyunin, *The inverse spectral problem for the modulus of a Hankel operator*, Algebra i Analiz **1** (1989) 54–66; translation in Leningrad Math. J. **1** (1990) 859–870.

167. Alexander Volberg, *Two remarks concerning the theorem of S. Axler, S.-Y. A. Chang and D. Sarason*, J. Operator Theory **7** (1982) 209–218.

168. Dragan Vukotić, *A note on the range of Toeplitz operators*, Integral Equations Operator Theory **50** (2004) 565–567.

169. Harold Widom, *Inversion of Toeplitz matrices. II*, Illinois J. Math. **4** (1960) 88–99.

170. Harold Widom, *On the spectrum of a Toeplitz operator*, Pacific J. Math. **14** (1964) 365–375.

171. Nina Zorboska, *Composition operators on weighted Dirichlet spaces*, Proc. Amer. Math. Soc. **126** (1998) 2013–2023.

172. Nina Zorboska, *Closed range essentially normal composition operators are normal*, Acta Sci. Math. (Szeged) **65** (1999) 287–292.

173. Nina Zorboska, *Cyclic composition operators on smooth weighted Hardy spaces*, Rocky Mountain J. Math. **29** (1999) 725–740.

Notation Index

Author Index

Subject Index

Graduate Texts in Mathematics

(continued from page ii)

121 LANG. Cyclotomic Fields I and II. Combined 2nd ed.
122 REMMERT. Theory of Complex Functions. *Readings in Mathematics*
123 EBBINGHAUS/HERMES et al. Numbers. *Readings in Mathematics*
124 DUBROVIN/FOMENKO/NOVIKOV. Modern Geometry—Methods and Applications Part III.
125 BERENSTEIN/GAY. Complex Variables: An Introduction.
126 BOREL. Linear Algebraic Groups. 2nd ed.
127 MASSEY. A Basic Course in Algebraic Topology.
128 RAUCH. Partial Differential Equations.
129 FULTON/HARRIS. Representation Theory: A First Course. *Readings in Mathematics*
130 DODSON/POSTON. Tensor Geometry.
131 LAM. A First Course in Noncommutative Rings. 2nd ed.
132 BEARDON. Iteration of Rational Functions.
133 HARRIS. Algebraic Geometry: A First Course.
134 ROMAN. Coding and Information Theory.
135 ROMAN. Advanced Linear Algebra. 2nd ed.
136 ADKINS/WEINTRAUB. Algebra: An Approach via Module Theory.
137 AXLER/BOURDON/RAMEY. Harmonic Function Theory. 2nd ed.
138 COHEN. A Course in Computational Algebraic Number Theory.
139 BREDON. Topology and Geometry.
140 AUBIN. Optima and Equilibria. An Introduction to Nonlinear Analysis.
141 BECKER/WEISPFENNING/KREDEL. Gröbner Bases. A Computational Approach to Commutative Algebra.
142 LANG. Real and Functional Analysis. 3rd ed.
143 DOOB. Measure Theory.
144 DENNIS/FARB. Noncommutative Algebra.
145 VICK. Homology Theory. An Introduction to Algebraic Topology. 2nd ed.
146 BRIDGES. Computability: A Mathematical Sketchbook.
147 ROSENBERG. Algebraic *K*-Theory and Its Applications.
148 ROTMAN. An Introduction to the Theory of Groups. 4th ed.
149 RATCLIFFE. Foundations of Hyperbolic Manifolds.
150 EISENBUD. Commutative Algebra with a View Toward Algebraic Geometry.
151 SILVERMAN. Advanced Topics in the Arithmetic of Elliptic Curves.
152 ZIEGLER. Lectures on Polytopes.
153 FULTON. Algebraic Topology: A First Course.
154 BROWN/PEARCY. An Introduction to Analysis.
155 KASSEL. Quantum Groups.
156 KECHRIS. Classical Descriptive Set Theory.
157 MALLIAVIN. Integration and Probability.
158 ROMAN. Field Theory.
159 CONWAY. Functions of One Complex Variable II.
160 LANG. Differential and Riemannian Manifolds.
161 BORWEIN/ERDÉLYI. Polynomials and Polynomial Inequalities.
162 ALPERIN/BELL. Groups and Representations.
163 DIXON/MORTIMER. Permutation Groups.
164 NATHANSON. Additive Number Theory: The Classical Bases.
165 NATHANSON. Additive Number Theory: Inverse Problems and the Geometry of Sumsets.
166 SHARPE. Differential Geometry: Cartan's Generalization of Klein's Erlangen Program.
167 MORANDI. Field and Galois Theory.
168 EWALD. Combinatorial Convexity and Algebraic Geometry.
169 BHATIA. Matrix Analysis.
170 BREDON. Sheaf Theory. 2nd ed.
171 PETERSEN. Riemannian Geometry. 2nd ed.
172 REMMERT. Classical Topics in Complex Function Theory.
173 DIESTEL. Graph Theory. 2nd ed.
174 BRIDGES. Foundations of Real and Abstract Analysis.
175 LICKORISH. An Introduction to Knot Theory.
176 LEE. Riemannian Manifolds.
177 NEWMAN. Analytic Number Theory.
178 CLARKE/LEDYAEV/ STERN/WOLENSKI. Nonsmooth Analysis and Control Theory.
179 DOUGLAS. Banach Algebra Techniques in Operator Theory. 2nd ed.
180 SRIVASTAVA. A Course on Borel Sets.

181 Kress. Numerical Analysis.
182 Walter. Ordinary Differential Equations.
183 Megginson. An Introduction to Banach Space Theory.
184 Bollobas. Modern Graph Theory.
185 Cox/Little/O'Shea. Using Algebraic Geometry. 2nd ed.
186 Ramakrishnan/Valenza. Fourier Analysis on Number Fields.
187 Harris/Morrison. Moduli of Curves.
188 Goldblatt. Lectures on the Hyperreals: An Introduction to Nonstandard Analysis.
189 Lam. Lectures on Modules and Rings.
190 Esmonde/Murty. Problems in Algebraic Number Theory. 2nd ed.
191 Lang. Fundamentals of Differential Geometry.
192 Hirsch/Lacombe. Elements of Functional Analysis.
193 Cohen. Advanced Topics in Computational Number Theory.
194 Engel/Nagel. One-Parameter Semigroups for Linear Evolution Equations.
195 Nathanson. Elementary Methods in Number Theory.
196 Osborne. Basic Homological Algebra.
197 Eisenbud/Harris. The Geometry of Schemes.
198 Robert. A Course in p-adic Analysis.
199 Hedenmalm/Korenblum/Zhu. Theory of Bergman Spaces.
200 Bao/Chern/Shen. An Introduction to Riemann–Finsler Geometry.
201 Hindry/Silverman. Diophantine Geometry: An Introduction.
202 Lee. Introduction to Topological Manifolds.
203 Sagan. The Symmetric Group: Representations, Combinatorial Algorithms, and Symmetric Functions.
204 Escofier. Galois Theory.
205 Félix/Halperin/Thomas. Rational Homotopy Theory. 2nd ed.
206 Murty. Problems in Analytic Number Theory.
Readings in Mathematics
207 Godsil/Royle. Algebraic Graph Theory.
208 Cheney. Analysis for Applied Mathematics.
209 Arveson. A Short Course on Spectral Theory.
210 Rosen. Number Theory in Function Fields.
211 Lang. Algebra. Revised 3rd ed.
212 Matoušek. Lectures on Discrete Geometry.
213 Fritzsche/Grauert. From Holomorphic Functions to Complex Manifolds.
214 Jost. Partial Differential Equations.
215 Goldschmidt. Algebraic Functions and Projective Curves.
216 D. Serre. Matrices: Theory and Applications.
217 Marker. Model Theory: An Introduction.
218 Lee. Introduction to Smooth Manifolds.
219 Maclachlan/Reid. The Arithmetic of Hyperbolic 3-Manifolds.
220 Nestruev. Smooth Manifolds and Observables.
221 Grünbaum. Convex Polytopes. 2nd ed.
222 Hall. Lie Groups, Lie Algebras, and Representations: An Elementary Introduction.
223 Vretblad. Fourier Analysis and Its Applications.
224 Walschap. Metric Structures in Differential Geometry.
225 Bump: Lie Groups.
226 Zhu. Spaces of Holomorphic Functions in the Unit Ball.
227 Miller/Sturmfels. Combinatorial Commutative Algebra.
228 Diamond/Shurman. A First Course in Modular Forms.
229 Eisenbud. The Geometry of Syzygies.
230 Stroock. An Introduction to Markov Processes.
231 Björner/Brenti. Combinatorics of Coxeter Groups.
232 Everest/Ward. An Introduction to Number Theory.
233 Albiac/Kalton. Topics in Banach Space Theory.
234 Jorgenson. Analysis and Probability.
235 Sepanski. Compact Lie Groups.
236 Garnett. Bounded Analytic Functions.
237 Martínez-Avendaño/Rosenthal. An Introduction to Operators on the Hardy-Hilbert Space.